T0142784

Internet of Things

Technology, Communications and Computing

More information about this series at http://www.springer.com/series/11636

Antonio Guerrieri · Valeria Loscri
Anna Rovella · Giancarlo Fortino
Editors

Management of Cyber Physical Objects in the Future Internet of Things

Methods, Architectures and Applications

Springer

Editors
Antonio Guerrieri
CNR—National Research Council of Italy
Institute for High Performance Computing
and Networking (ICAR)
Rende, CS
Italy

Valeria Loscri
Parc scientifique de la Haute Borne
Inria Lille-Nord Europe
Villeneuve d'Ascq
France

Anna Rovella
University of Calabria
Rende, CS
Italy

Giancarlo Fortino
DIMES
University of Calabria
Rende, CS
Italy

ISSN 2199-1073 ISSN 2199-1081 (electronic)
Internet of Things
ISBN 978-3-319-80030-1 ISBN 978-3-319-26869-9 (eBook)
DOI 10.1007/978-3-319-26869-9

Springer Cham Heidelberg New York Dordrecht London

Softcover reprint of the hardcover 1st edition 2016

Printed on acid-free paper

Springer International Publishing AG Switzerland is part of Springer Science+Business Media (www.springer.com)

Preface

The emerging generation of Cyber Physical Objects (CPOs) brings new powers opening novel and revolutionary applications and new areas of innovation. The Cyber Physical Systems (CPS) constituted by the CPOs enable fusion between the physical and virtual worlds realizing the totally and globally connected scenario known as Internet of Things (IoT).

In the past decade, Cyber Physical Systems and consequently Cyber Physical Objects as their basic units and IoT as their reference scenario have become one of the ICT priorities.

The range of potential application domains is broad and embedded hardware and software systems could expand functionalities of household appliances, vehicles, aircraft, medical applications, digital libraries, etc. The interconnection of CPOs, through a virtual environment featured with globally networked services, is opening towards innovative business platforms.

The pervasive "features" of the emerging CPOs perfectly match and embrace the philosophy of the future "Internet of Things" (IoT), as the vision that nearly everything is connected to the Next-Generation Internet. This emerging paradigm is demanding novel methods and revolutionary architectures to respond to innovative applications in several daily scenarios. Methods for engineering CPSs that will be able to respond in real-time to dynamic and complex situations while preserving control reliability, safety, and keeping security and privacy need to be designed ad hoc.

Based on these premises, it is easily arguable that new issues and challenges arising with the IoT paradigm and the definition of CPSs need to be identified and timely addressed.

The main objective of this book is to explore multidisciplinary and recent advances in terms of methods, architectures, and identify novel applications deriving from the CPOs as main building blocks of the Internet of Things paradigm.

The book is structured into eight chapters focused on the above-mentioned topics and providing novel and cutting-edge contributions for the Next-Generation IoT systems. A short introduction to the chapters is provided below.

Chapter "Cyber-Physical Systems: Opportunities, Challenges and (Some) Solutions," by Peter Marwedel and Michael Engel, discusses the integration of information and computation technologies (ICT) with real and physical objects. It motivates the work in the Cyber Physical Systems area by presenting the large set of opportunities resulting from this integration. However, this requires coping with a number of challenges that are also discussed. The chapter comprises solutions demonstrating that it is feasible to address the challenges and find solutions, even though a major amount of additional work is required.

Chapter "Cyber-Physical Objects as Key Elements for a Smart Cyber-City," by Riccardo Petrolo, Valeria Loscri, and Nathalie Mitton, investigates the Cyber Physical Systems and their Cyber Physical Objects as key units, in the context of Smart City concept. It surveys the Smart City vision, providing information on the main requirements, the open challenges, and highlighting the benefits; it also browses the European Commission initiatives for Smart Cities and some pilot projects that are under development.

Chapter "Structuring Communications for Mobile Cyber-Physical Systems," by Luis Almeida, Frederico Santos, and Luis Oliveira, addresses the problem of developing applications for, and controlling, Mobile Cyber Physical Systems composed of a team of mobile autonomous agents. It focuses on the communications and middleware platform on which most cooperative behaviors rely. For such purpose, a platform that allows sharing state among team members, while abstracting away communication, has been developed. The platform relies on a shared memory middleware that extends the traditional Blackboard concept with local data proxies that also include data age information.

Chapter "ANIMO, Framework to Simplify the Real-Time Distributed Communication," by Y. Rodríguez, C. Alejo, I. Alejo, and A. Viguria, presents ANIMO, an IoT communication framework developed for interconnecting multi-systems based on the Data Distribution Service (DDS). ANIMO facilitates the integration of DDS in an application and the interoperability between the different data types of the Cooperating (Cyber Physical) Objects with the great feature of real-time. This chapter explains the complete architecture of the ANIMO framework, its diversity of possibilities, and two principal works where it has been applied.

Chapter "SERAPH: Service Allocation Algorithm for the Execution of Multiple Applications in Heterogeneous Shared Sensor and Actuator Networks," by Claudio Miceli de Farias, Wei Li, Flávia C. Delicato, Luci Pirmez, Paulo Pires, and Albert Y. Zomaya, proposes an adaptive algorithm (called SERAPH) to select and allocate what they call services, for multiple applications in a heterogeneous sensor platform. They regard the applications as a set of primitive services, which each node in a Shared Sensor and Actuator Network is able to exhibit. The authors compare their algorithm with three other approaches in terms of network lifetime, allocations successful rate, and loss rate and delay.

Chapter "A Smart Platform for Large-Scale Cyber-Physical Systems," by Andrea Giordano, Giandomenico Spezzano, and Andrea Vinci, introduces Rainbow, an architecture designed to address Cyber Physical Systems issues.

Rainbow hides heterogeneity by providing a Virtual Object concept, and addresses the distributed nature of Cyber Physical Systems introducing a distributed multi-agent system on top of the physical object. Rainbow aims to get the computation close to the sources of information and tackles the dynamic adaptivity requirements of Cyber Physical Systems by using Swarm Intelligence algorithms.

Chapter "Towards Cyberphysical Digital Libraries: Integrating IoT Smart Objects into Digital Libraries," by Giancarlo Fortino, Anna Rovella, Wilma Russo, and Claudio Savaglio, proposes an approach for the inclusion of Smart Objects into Digital Libraries. The inclusion is based on a metadata model for Smart Objects purposely defined to fully characterize all the Smart Object properties (both physical and cyber) as well as their interactions with other human, digital, and cyber physical actors. The approach has also been exemplified through a case study concerning a smart office environment.

Chapter "Cooperation of Smart Objects and Urban Operators for Smart City Applications," by Simona Citrigno, Sabrina Graziano, and Domenico Saccà, describes a set of software tools and intelligent platforms for collecting, representing, managing, and exploiting data and information gathered from sensors and devices deployed in the territory. Tools and platforms are integrated into a complex smart environment that provides advanced services to citizens and operators for environmental monitoring, urban mobility, and emergency management.

We would like to thank all the book contributors, the anonymous reviewers, and Ravi Vengadachalam from Springer for his support and work during the publication process.

Antonio Guerrieri
Valeria Loscri
Anna Rovella
Giancarlo Fortino

Contents

Cyber-Physical Systems: Opportunities, Challenges and (Some) Solutions

Peter Marwedel and Michael Engel

Abstract The notion of Cyber-Physical Systems (CPS) has recently been introduced. The term describes the integration of information and computation technologies (ICT) with real, physical objects. In this chapter, we motivate work in this new area by presenting the large set of opportunities resulting from this integration. However, this requires coping with a number of challenges which we do also include in this chapter. The final main section of this chapter comprises solutions which demonstrate that it is feasible to address the challenges and find solutions, even though a major amount of additional work is required.

1 Introduction

Until the late eighties of the last century, information processing was associated with large mainframe computers and huge tape drives. During the nineties, this shifted towards information processing being associated with personal computers, PCs. The trend towards miniaturization continues and the majority of information processing devices will be small portable computers integrated into larger products. It is obvious that many technical products have to be technologically advanced to find customers' interest. Cars, cameras, TV sets, mobile phones, etc. can hardly be sold any more unless they come with built-in computers. Following the success of information and communication technologies (ICT) for office and work flow applications, ICT components embedded into enclosing products are considered to be the most important application area of ICT during the coming years.

P. Marwedel (✉) · M. Engel
TU Dortmund, Dortmund, Germany
e-mail: peter.marwedel@tu-dortmund.de

M. Engel
e-mail: michael.engel@tu-dortmund.de

A. Guerrieri et al. (eds.), *Management of Cyber Physical Objects in the Future Internet of Things*, Internet of Things,
DOI 10.1007/978-3-319-26869-9_1

The presence of ICT in these products will be less obvious than for the PC. In the future, ICT devices will be embedded into products such that, for most users, they will actually become invisible. Due to this trend, Weiser [54] introduced the term "disappearing computer". The fact that ICT components (computers) are embedded into systems led to the term "embedded systems", which can be defined as follows [30]:

Definition Embedded systems are information processing systems embedded into enclosing products.

Examples include embedded systems in cars, trains, planes, telecommunication or fabrication equipment. Such systems share a large number of common characteristics, including real-time constraints and dependability as well as efficiency requirements. For such systems, the link to physics and physical systems is important. This link has been emphasized by Lee [26]:

> Embedded software is software integrated with physical processes. The technical problem is managing time and concurrency in computational systems.

This citation could be extended into a definition of "embedded systems" by just replacing "software" by "system". However, the link to physics has recently been stressed even more by the introduction of the term "cyber-physical systems" (CPS or "cy-phy" systems for short). Cy-phy systems can be defined as follows:

Definition "*Cyber-Physical Systems (CPS) are integrations of computation and physical processes*" [27].

The new term emphasizes the link to physical quantities such as time, energy and space. Emphasizing this link makes sense, since it is frequently ignored in a world of applications running on PCs. Further discussions of this topic led to the funding of research on CPS by the National Science Foundation (NSF) in the US. In their call for proposals, NSF is describing CPS as follows:

> Cyber-physical systems (CPS) are engineered systems that are built from and depend upon the synergy of computational and physical components. Emerging CPS will be coordinated, distributed, and connected, and must be robust and responsive [37].

The definition also stresses the importance of distribution and networking and, in this way, goes beyond an interface between physical and ICT components. This emphasis is also present in the description of CPS by the German Akatech organization, which followed the NSF proposal. According to Akatech,

CPS "*represent networked, software-intensive embedded systems in a control loop, provide networked and distributed services*" [3].

In this case, control loops are also explicitly mentioned. Control loops are considered in several of the proposals for research on CPS, including proposals supported by NSF. In its report, Akatech explicitly presents four application areas of CPS: mobility, health, energy, and production. In Germany, innovation in production is frequently associated with the term "Industrie 4.0" [19], referring to the fourth generation of factory design, after the introduction of steam engines, electricity and numerical control. Industrie 4.0 links design phases, logistics and production. In principle, each customer could have his or her own custom designed version of a product.

More recently, the European Commission has also started a program supporting the design of CPS. In a call for the current eighth funding framework, CPS is explained as follows:

> Cyber-Physical Systems (CPS) refer to next generation embedded ICT systems that are interconnected and collaborating including through the Internet of things, and providing citizens and businesses with a wide range of innovative applications and services. These are the ICT systems increasingly embedded in all types of artifacts making "smarter", more intelligent, more energy-efficient and more comfortable our transport systems, cars, factories, hospitals, offices, homes, cities and personal devices. Often endowed with control, monitoring and data gathering functions, CPS need to comply with essential requirements like safety, privacy, security and near-zero power consumption as well as size, usability and adaptability constraints [14].

Distribution, connectedness and control loops are highlighted as well. Also, a number of expected application areas and properties are included.

In a way, the term CPS describes many opportunities arising from the cooperation of ICT and physical components. We do expect that the term will be used for many new projects linked to ICT. The precise boundaries are blurring. We could discuss how many new aspects are considered in CPS which were not already discussed in the context of embedded systems. However, this can be a rather fruitless discussion. Therefore, we prefer to look forward and discuss opportunities arising from CPS as they were introduced above. We will present such opportunities in the next section. Section 3 is devoted to challenges for implementing CPS. Section 4 highlights some of the solutions for these challenges. This chapter will close with a conclusion in Sect. 5.

2 Opportunities

There are many opportunities resulting from the integration of computing and the physical environment. In the following, we would like to demonstrate the large set of opportunities by listing some of the prominent sample areas.

2.1 Transportation and Logistics

The introduction of this chapter does already highlight some the potential applications of CPS technologies in transportation.

More precisely, we can point to the fact that—in industrialized countries at least—no contemporary car can be sold unless the car provides a good amount of ICT components. Such components include engine control, comfort features and safety features such as electronic stability programs (ESP). In order to meet environmental standards, ICT technologies are required. Parking assistants are being offered on a regular basis. There is huge progress in autonomous driving [39].

For rail-based systems, ICT components are added as well. High-speed trains are feasible only due to ICT components. The next generation safety standards are also based on an increased amount of ICT.

For avionics, the use of ICT components is also increasing. Even pilot-less flying is possible. In fact, it is already happening with drones and is expected to gradually become reality for airplanes as well, starting with freight planes [47].

The fourth means of transportation, ships, is also using an increasing amount of ICT components. For example, these can help finding the best route or even enable autonomous operation of vessels [12].

Very much linked to transportation, we have logistics, comprising the optimization of supply chains and just-in-time delivery. Efficient parcel delivery is of extreme importance, due to the increasing amount of customers ordering through the Internet. Storage and retrieval of goods can also be supported by ICT components [4].

2.2 Fabrication/Production

Closely related to logistics, we have fabrication. There are many opportunities for optimization of fabrication by using ICT. Fabrication is actually one of the application areas of CPS mentioned in the Akatech report [3]. The large amount of opportunities led to the introduction of the term “Industrie 4.0” in Germany [19]. Expectations for innovation in this area are high.

2.3 Smart Home

There are many opportunities for improving life at home with respect to various metrics. For example, we can improve the level of safety, energy efficiency and comfort. In particular, we can support elderly people. Ambient assisted living is one of the targets here. Connecting the various devices in a home can improve energy efficiency and safety.

The so-called zero-energy building is one of the special cases. Such a building is supposed to generate as much energy as it is consuming, at least on average. In order to turn this vision into reality, we can use solar cells, smart ventilation and control of blinds, as well as energy-efficient heating and lighting. The consumption of energy has to be adjusted to the actual use of features. For example, air conditioning of empty rooms can be reduced. In some cases (like for washing machines and charging of batteries), consumption times can be adjusted to times of availability.

2.4 Health

The health sector comprises a huge amount of different applications of this integration.

This starts with CPS-supported diagnosis. We can design new, ICT-enabled sensors (later in this chapter, we will be presenting one). We can have advanced data and risk analysis techniques (they are studied in our collaborative research center SFB 876 [34]). Supply chains can be monitored in order to identify sources of problems.

Once a diagnosis has been completed, therapies can be supported as well. Personalized medication is within reach: with this concept, every person will receive just the right amount of medication. Result monitoring is also becoming feasible due to the use of sensors. Handicapped patients can be supported in an appropriate way.

Finally, patient information systems can provide detailed information about a patient, in this way avoiding redundant analysis. The health sector is also one of the four sectors mentioned in the Akatech report.

2.5 Further Opportunities

There are many more opportunities. They cannot be covered in detail. Therefore, we provide only a list:

- Structural health monitoring: It is feasible and beneficial to monitor the structural health of artificial and natural objects. For example, we can try to predict the risk of falling rocks or collapsing bridges [36].

- Disaster recovery: ICT can help to organize rescue operations following disasters. Communication is a key in such a scenario.
- Public safety: There is an increased demand on public safety. Places of increased vulnerability are frequently monitored by cameras. Research efforts aim at the automatic identification of periods of higher risk. Privacy concerns are preventing the fast grow rate which we could otherwise expect in this area.
- Scientific experiments: Most of the experiments in physics require an analysis using ICT components. By definition, this is a classical case of CPS.
- Robotics is one of the roots of CPS.
- Energy grids must become smart in order to cope with the challenges of renewable energy sources, in particular their fluctuating generation of power.
- We include telecommunication here as well, since it is a key ingredient for the exchange of information between connected components.
- Finally, we include consumer electronics, since it shares many features with other types of CPS, like the limited amount of resources, including energy and the presence of real-time constraints. However, this inclusion is debatable.

A few examples from the catalog of projects supported by NSF prove the existance of a wide scope and a wide range of opportunities:

1. One project considers applications of oceanographic research [7].
2. A second project concerns a programmable lab on a chip, with applications in life-sciences [6].
3. A third project considers validation and monitoring in robotic surgery systems [8].
4. Air traffic management is studied in another project [2].

2.6 *Summary of Opportunities*

Summarizing the opportunities mentioned above, we can conclude that there is an abundant amount of opportunities for the integration of ICT components and the physical environment. In a way, the term CPS comprises most of the coming applications of ICT, with the exception of traditional office and compute center applications.

3 Challenges

Unfortunately, there are many challenges on the way to the bright future of ubiquitous support by CPS in our environment.

3.1 Security

We believe, that security will be the number one challenge in the coming years. Several examples demonstrate the importance of caring about security:

- In 2014, the Penemon Institute estimated the annual worldwide cybercrime costs at $160 billion [42].
- Cyber attacks such as Stuxnet [18] found world-wide attention.
- Cyber terrorism has become a real threat to our infrastructures. For example, the potential impact on our power distribution is being considered.
- Cyber war has already become reality. The use of drones can be seen as an early example of this. Consequences of the new options are largely unknown.

The key feature of CPS, global connectedness, is actually also the key reason for security threats. The fact that "everything" is connected is also enabling adversaries to affect "everything" in a malicious attack. A large amount of effort is required in order to fight against such attacks. No perfect solution seems to be feasible, unless we give up on the idea of global connectedness. In fact, highly secure islands of very sensitive information are typically not connected at all and the only guaranteed safe way of preventing information leaks is not to capture the information. So, as long as we want to stay connected and we want to capture information, we have to invest into security-improving solutions.

3.2 Safety

The fact that CPS are tightly integrated into the physical environment is also a potential hazard. Errors in the information processing part of the system can spread into the physical parts and can be dangerous to people, animals or our environment. Obvious examples include cars (especially self-driving cars), airplanes, trains, production sites and devices in the health sector. Threats to safety can be diverse: they could result from improper specifications, from design errors, from the failure to provide the expected operating condition and also from hardware failures. Safety needs to have far-reaching consequences: they may imply the necessity to employ formal verification techniques such as model checking or the use of special programming languages like synchronous languages [44].

3.3 Reliability

We denote by reliability the probability of providing the expected service at run-time. Failures are cases of not providing the service at run-time. The underlying reasons for failures can be diverse: there may be design errors, hardware errors or environmental conditions outside the specified range. Reliability requirements may come in the form of ceilings for tolerated failure rates. For example, one might specify a ceiling of 10^{-9} failures per hour for safety-critical systems.

Failure rates of actual hardware devices may be orders of magnitude larger than the mentioned failure rate. For example, Triquint published failures rate of up to almost 10^{-6} per hour for late phases in a circuit's life [52]. For shrinking semiconductor device dimensions, these values may even increase. This means that fault tolerance technologies must be employed to reach the required failure rates. Significant research into new fault tolerance technologies is required.

Overall, we aim at dependable systems, that is, systems with a low probability of service failures, even when they are not used under the conditions laid down in the specification.

3.4 *Timing Predictability*

Time is one of the most important physical quantities. Linking information processing to physical systems also implies that the timing behavior of information processing becomes extremely important. In fact, many CPS are real-time systems. For such systems, it is necessary to react within a time interval dictated by the environment. The consequences of this requirement are frequently underestimated. Many systems are designed such that they provide a sufficiently fast response time in most of the cases, but may fail to do so in some others. For example, there may be communication retries, access conflicts, accesses to slow members of the memory hierarchy, etc. which really slow down a system, especially if worst case behaviors of components are combined. Desktop- and server-based computing have very much been influenced by the absence of the need to provide real-time guarantees. For CPS, this needs to be changed. Changes may affect the entire hardware/software stack, from the design of caches via the design of access to shared resources in multi-core systems to the timing of operating system calls and communication protocols.

3.5 *Energy Efficiency*

Many CPS are mobile systems or have to be operated in areas where only a limited amount of electrical energy is available. As a result, such systems either have to generate electrical energy themselves (energy harvesting) or they have to use batteries. In both cases, energy is a very scarce resource and must be used carefully.

As an example, we would like to use the smart "InBin" box for applications in logistics. The box is designed by the IML Fraunhofer Institute at Dortmund [20]. It is aware of its content and can communicate its content and location to the local computer network. InBin contains two processors for this purpose. The box uses energy harvesting by solar cells. Energy consumption is low enough to allow an operation of the box even in rather dark storage facilities. This design clearly demonstrates the challenges which may exist in the design of low power systems.

3.6 Human and Social Issues

Presently available ICT systems already have a large impact on society. For example, some types of businesses like movie theaters, newspapers, book publishers, travel agents, and shops are severely affected by the availability of the Internet. The role of employees is being changed and there is even an impact on the way governments are run. Future CPS have an influence even on a larger number of human activities. The overall impact on the human society is largely unknown, since social sciences can hardly cope with the speed of changes. It will be a challenge to ensure that, overall, the use of CPS will be to the benefit of the majority of the society as well as to our environment and not just to the benefit of a few. Furthermore, there may be cases in which CPS technology will not be accepted by society, including cases when it would be beneficial. In other cases, technology might be accepted even when it is dangerous to do so. Hence, education and public discussion have to go hand-in-hand with the development of new technology.

3.7 Legal Issues

Connecting a large number of components together implies that we are connecting components of different manufacturers. Each of them may be operated by a different legal entity. A natural question arises: Who owns the data? This question is actually very difficult to answer. Problems do already exist with social networks. Typically, each social network operator claims to own the data. If more legal entities are involved, this approach becomes unmanageable.

Liability is another important issue. In the case of some damage, one has to identify the legal entity which is liable. In case several legal entities are involved, it is difficult to figure out, who should compensate the damage. Unsolved legal issues could actually prevent useful CPS technology from being implemented.

3.8 Dynamism

Desktop systems are essentially remaining at a fixed location. Their network interface is hardly ever changing. This is different for CPS which are being moved around. The network connection is changing frequently. Different networks based on wireless LAN, Bluetooth®, mobile phone standards, etc. can be used. Each may have different connection speeds and costs. The connection quality may be changing over time. Network delays may vary. Due to this, network connections have to provide a high level of fault tolerance.

The amount of available energy, available external devices, and the computational load may also change over time. All of this means that operating conditions may be highly dynamic.

3.9 Heterogeneity

Connecting many different components also implies that we are connecting components of different types. Some may be analog components, whereas others may be digital components. Components may be designed for different communication protocols, by different companies, or for regional variants. Achieving interoperability despite this heterogeneity is a challenge.

3.10 Multidisciplinary Nature

Obviously, the design of CPS requires knowledge from various domains, including physics and computer science. There are also many aspects which require knowledge in electrical engineering. For example, communication engineering is typically linked to electrical engineering. Furthermore, knowledge from disciplines like medicine, mechanical engineering, biology, chemistry, and statistics may be necessary for many applications. As mentioned earlier, some issues require knowledge in social sciences or law. Due to tight ceilings imposed on student workloads and credit points, it is impossible to incorporate all of this into one curriculum. Hence, it is necessary to design CPS educational programs such that some core knowledge is guaranteed while the ability to lead discussions with specialists for the other areas must be available as well.

3.11 Verification Problem

Verification of digital systems has typically tried to find an answer to a Boolean question: Is this system a correct implementation of my specification or not? In the context of physical quantities, it is not appropriate to phrase the verification problem in this way. Many of the physical quantities are analog quantities. Measurement of analog quantities comes with inevitable uncertainty, at least due to the Heisenberg uncertainty principle. Hence, we are unable to demonstrate that a real CPS is an exact implementation of the specification. We have to replace

Boolean verification by a kind of Fuzzy verification. Fuzzy verification provides a level of confidence $\ell, 0 \leq \ell \leq 1$ for a correct implementation [41].

3.12 Inappropriate Number Systems

Many physical quantities are represented as a pair (*n*, *u*) consisting of a real number *n* and a unit *u*. There are infinitely many real numbers. In fact, in between any two such numbers, there are infinitely many more numbers. This is completely different within the digital information processing part of CPS: in practice, we can use only a finite set of symbols. So, the majority of pairs cannot be represented. Floating or fixed point numbers are used as approximations. Absolute resolution frequently depends on the absolute value. This leads to the question: do these approximations actually represent physical reality? If the relationship between physical quantities and observed physical phenomena were completely chaotic, discrete approximations of physical systems would be of limited value. Fortunately, monotonic behavior can be observed between two discrete approximations if the difference between these two approximations is small enough. Nevertheless, this mismatch of number systems results in challenges for the simulation of CPS [51].

3.13 Selection of Sampling Intervals

Information processing in digital systems is usually based on periodic sampling of analog signals. Sample periods are typically selected by rules of thumb. For example, physical systems can be approximated with equivalent electrical circuits (e.g. for thermal modeling) comprising time constants of RC-combinations. Sample periods are then selected as the shortest time constant divided by some small factor. Noise may have a negative impact if the sampling period is made much shorter, which therefore is typically avoided. This approach has worked well for many applications. However, this approach may result in a large amount of actions like computations and communications which are not needed for a safe and stable operation. For example, if the values of physical quantities and corresponding sensor signals remain essentially unchanged, no new sampling may be required. Sampling could focus on cases of rapidly changing signals. At least, this could result in techniques for energy-efficient control loops. In such a scenario, the selection of the time until the next sampling can become an art by itself. Machine learning can be used for this [50].

3.14 Zeno Behavior

In physics, there are cases for which, under a certain model, an unbounded number of events or changes can happen in a finite amount of time. For example, consider the case of an elastic ball which gets dropped from some height. Suppose it bounces on the floor. We assume that the bouncing happens at height $y = 0$ due to either approximating the ball as a point mass or due to shifting y by the radius of the ball. Suppose there is some damping by a factor of γ, resulting in a reflection of the ball not reaching exactly the same maximum height before falling down again. This process will repeat and for each reflection, a lower height is reached. Also, the time between reflections will get shorter and shorter. Equations in Fig. 1 describe a single bouncing at $t = 0$.

This process will continue indefinitely unless some other effect (e.g. a sticky behavior) is assumed. Now consider a time interval T. Can we guarantee an upper bound on the number of reflections within this interval T? No, we cannot, since the interval between reflections is getting shorter and shorter.

The bouncing ball is a special instance of Zeno behavior. A system exhibits Zeno behavior if we cannot guarantee a finite number of events (e.g. reflections) in a finite amount of time [35].

The simulation of Zeno behavior on a digital computer is not trivial and requires special care, including approaches taking the limited resolution of number systems into account.

3.15 More Challenges

There are even more challenges for the design of CPS. We briefly mention some of them: they include

- sensitivity of CPS to radiation (including electromagnetic radiation and radioactivity),

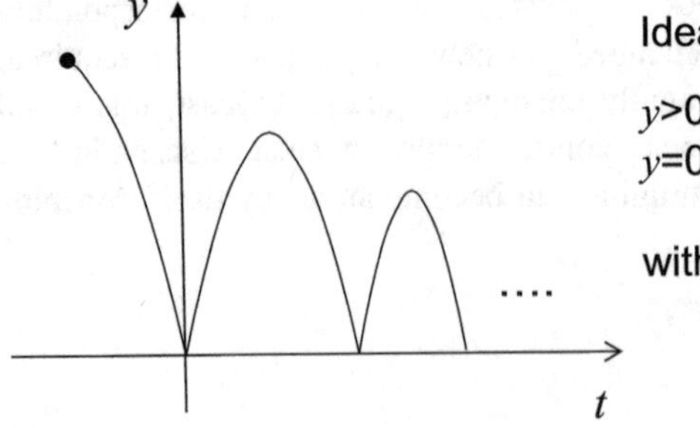

Fig. 1 Bouncing ball with Zeno effect

- large amounts of data ("big data") generated in some scenarios,
- exactness issues (how exact is the result?),
- cost issues (of course, we would like to have cheap CPS),
- size issues (the CPS has to fit into the available space),
- weight issues (the CPS must be carried somehow),
- environmental friendliness,
- user friendliness,
- extensibility.

4 (Some) Solutions

Due to the large number of challenges, there may be doubts as to whether or not CPS can be implemented at all. Of course, compromises have to be found. We would like to demonstrate, that some solutions exist, even though they are typically not addressing all challenges. Our focus will be on contributions made by our own group, except for the very first example.

4.1 Modelica

The Modelica simulation system is our first example. This system shows that simulation of physical systems, despite all the challenges, can be performed. Modelica starts from a schematic model of the components (e.g. pumps, reservoirs etc.) and their connections. Components consist of connected sub-components or are described by equations. Figure 2 provides an example.[1]

This schematic model is transformed into a set of differential equations. These equations are then fed into one of the tools for solving differential equations, e.g. Matlab. More information is available on the Modelica web site [40].

4.2 Energy-Efficient Bio-virus Detection

A special combination of an advanced sensor and advanced information processing makes up our second example. In this case, the aim is to be able to detect (bio-) viruses fast. A new sensor (see Fig. 3) is the key component.

The sensor comprises a prism. On one side of the triangular prism, there is a very thin layer of gold. Laser light is entering the prism from a second side. The light

[1]Figure taken from Modelica Overview by Otter and Winkler [40] under the Creative Commons Attribution-Share Alike license.

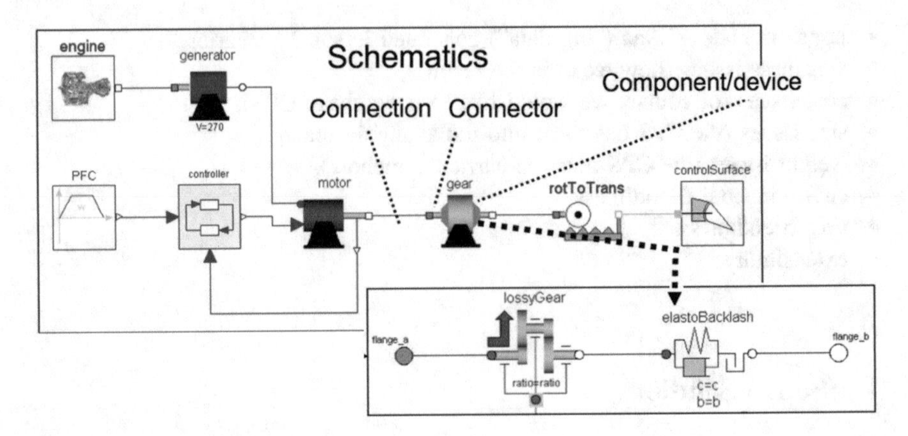

Fig. 2 Specification in Modelica

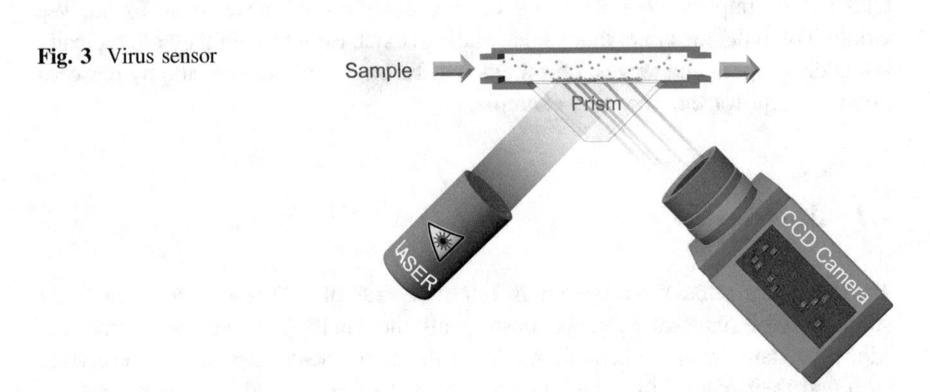

Fig. 3 Virus sensor

gets reflected at the gold layer and the reflected light leaves the prism through the third side. Assuming a correct adjustment of the beams, the reflected light is detected by a video camera. Viruses can be detected in liquids or gasses potentially flowing along the thin gold layer, at the layer's side opposite to the prism. Viruses can get stuck at the gold layer, by binding to the antibodies on top of the gold layer. One could have the impression that this should have no impact on the amount of light seen by the camera since the gold layer is still between the viruses and the light beam and since viruses are typically smaller than the wavelengths of light. However, the gold layer is extremely thin and viruses are causing a so-called resonance effect which changes reflectivity in a region larger than the size of the virus. As a result, even single viruses which get stuck on the gold layer will cause an observable change in the intensity of the reflected light. This change can be detected and hence, single viruses can be detected. This has potential applications in the fast recognition of viruses, for example at hospitals.

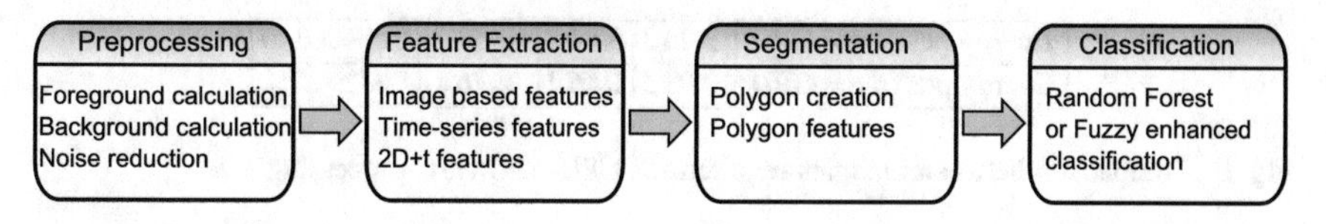

Fig. 4 Image processing pipeline

Due to the small dimensions involved and because the changes in reflectivity caused by a single virus are very small, images captured by the camera contain quite some noise. Sophisticated image analysis is required in order to discriminate real viruses from noise. Figure 4 shows the image processing pipeline [28] we are currently using.

The integration of a physical sensor with this information processing pipeline highlights the fact that we are dealing with a cyber-physical system.

In order to enable real-time image processing, we are using graphics processors (GPUs). Table 1 shows analysis rates that can be achieved for different CPUs and GPUs for different resolutions [28].

The rates are capped at the screen refresh rate of 60 Hz. Real-time image processing for a Intel i7-2600 desktop CPU is only feasible for a resolution below 1024 × 128, while it is not feasible for the tested mobile Intel i7-620M CPU. For the NVIDIA 3100M mobile GPU real-time image processing is possible for resolutions up to 1024 × 256. GPUs with a higher performance, like the NVIDIA GTX 480 desktop GPU, would allow us to increase the resolution beyond the one we are currently aiming at.

While GPUs offer the required performance, there may still be the question whether or not the use of GPUs is energy effcient. In order to answer that question, we measured the energy consumption of a NVIDIA 3100M GPU card, compared to an Intel i7-620M CPU. Figure 5 shows the results for three different resolutions. We are able to demonstrate that we need only 27 % of the energy that is required for a CPU-based solution. In this way, we are able to demonstrate that we can improve the energy efficiency of a cyber-physical system.

Further optimizations could be achieved by porting the Pamono code base to the Odroid XU3 [22], an embedded platform utilizing a state-of-the-art heterogeneous embedded eight-core ARM BIG.little [43] processsor with an integrated

Table 1 Analysis rates attained for different image sizes and different devices

Frames per second (fps)		Image size in pixels			
		1024 × 128	1024 × 256	1024 × 512	1024 × 1024
Desktop	CPU: i7-2600	23	20.1	16.1	11.5
	GPU: GTX 560 Ti	60	60	60	26.8
	GPU: GTX 480	60	60	60	40.3
Laptop	CPU: i7-620M	16	8.5	4.6	2.4
	GPU: 3100M	60	32.6	16	7.8

Energy per frame CPU	3.26 J	5.84 J	10.52 J	Reduced to
Energy per frame GPU	0.93 J	1.56 J	2.76 J	avg. 27%

Fig. 5 Comparison between energies required for CPU- and GPU-processing

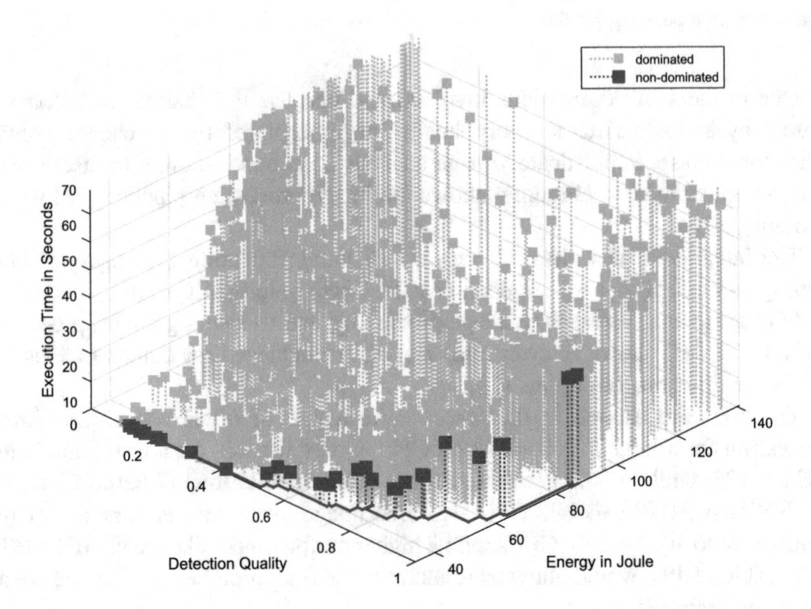

Fig. 6 Pareto-optimal solutions for the Pamono optimization on the Odroid platform

OpenCL-capable Mali T628 GPU [38]. Using a multi-criterial hardware in the loop approach to generate and evaluate different configurations of the detection software generated by a genetical algorithm, we were able to run the Pamono software in real-time on this embedded platform while saving up to 76 % of energy compared to an unoptimized baseline. Some of the Pareto-optimal solutions for an optimization of runtime versus energy and detection quality are shown in Fig. 6.

4.3 Worst Case Execution Time Aware Compilation

We did already mention the fact that CPS require paying attention to the timing behavior. In particular, we need to make sure that computations are completed within the time interval that is available. This leads us to having a closer look at execution times of programs. If we execute a program, we will in general observe an execution time which depends on input values. If, hypothetically, we run the program with all possible input patterns, we will—in the general case—observe some execution times more frequently than others. Overall, there will be a distribution of execution times (see Fig. 7).

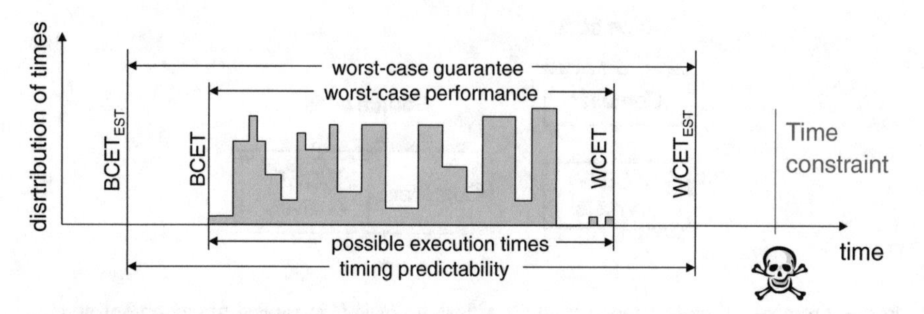

Fig. 7 Notion of worst-case execution times

Among all execution times, there will be one, which is the largest. We call it the "worst case execution time" (*WCET*). Also, there will be one execution time which is the smallest. We call it the "best case execution time" (*BCET*). Unfortunately, the *WCET* and the *BCET* can typically not be computed exactly, since the number of input patterns times the number of initial states is such a big number that complete simulation, with all possible combinations, is not feasible. Hence, results of such "dynamic" techniques have to be considered with care. Static techniques are an alternative to dynamic techniques. For static techniques, programs are evaluated off-line, using, for example, abstract interpretation [10]. Abstract interpretation is safe by design. However, in order to use abstract interpretation, safe abstractions are required. For example, marginal cases leading to long execution times may be modeled as if they were feasible, even though they are not. As a result, abstract interpretation provides a safe upper bound on the execution time, which we call "estimated worst case execution time" ($WCET_{EST}$). We require that $WCET_{EST}$ is a safe approximation of *WCET* and that it is a tight estimate (the difference between $WCET_{EST}$ and *WCET* should be small). In a similar way, we can define the "estimated best case execution time" ($BCET_{EST}$).

Now, suppose that there is a deadline *D* until which our computations must be completed. We will then have to check whether $WCET_{EST} \leq D$ holds or not. If this relation does not hold, we have to change some detail of the design, potentially also details such as compiler options. Next, we will have to rerun the check. Such "trial-and-error" iterations can take quite some time and it would be nice if they could be avoided.

One possibility for at least reducing the number of iterations can be detected by revisiting the standard software production. In the standard approach, there is some specification from which the software written in a high-level language is produced. This software is then compiled. During compilation, the compiler will try to "optimize" generated code, but the cost function is unclear and not considered except for a few compilation steps. If the software production is done carefully, the resulting code is then fed into a tool for static *WCET* estimation. If the deadline is violated, we have to go back to one of the earlier design steps.

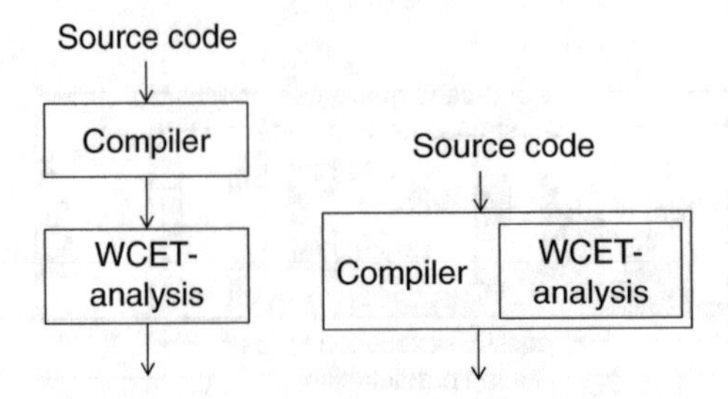

Fig. 8 From post-code-generation timing verification towards integrated timing optimization

Obviously, there is hope that some of these iterations can be avoided, if the compiler is already aware of the $WCET_{EST}$ as the objective to be minimized. It could then favor the generation of code which for worst case input data is faster than some other code. Figure 8 visualizes the change from post-compilation computation of $WCET_{EST}$ to compilation with integrated $WCET_{EST}$ minimization.

In order to minimize $WCET_{EST}$, we have created the compiler structure shown in Fig. 9.

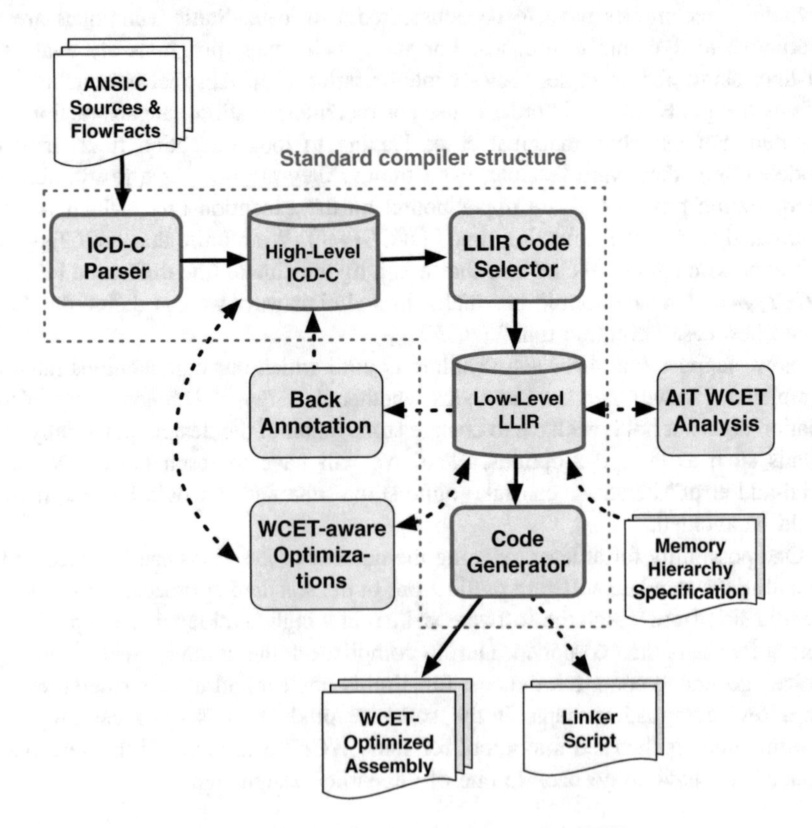

Fig. 9 Structure of the worst-case execution-time aware compiler WCC

Components shown within the dashed outline are standard compiler components. These are designed to read in C source code, to represent this code in an internal high-level format, to map high-level operations to machine operations and to represent these at the corresponding lower level (LLIR). Optimizations can be performed at both internal representations. The essential change compared to a standard compiler is the incorporation of *WCET*-analysis as well as the addition of back-annotation and *WCET*-aware optimization. Towards this end, we added a link to the commercial aiT *WCET* analyzer [1]. This link provides the compiler with $WCET_{EST}$ estimates, so that optimizations can, among various different codes, select the one that has the smallest value of $WCET_{EST}$. The resulting compiler is the experimental “worst case execution time aware compiler” WCC. In order to speed up the optimization and in order to improve the flexibility, later versions of WCC use their own local abstract interpretation package.

This integration of *WCET*-estimation enables a fresh new look at almost all known compiler optimizations: we can analyze their potential for reducing $WCET_{EST}$. Much hope relied on *WCET*-aware inlining: inlining is performed at a high-level, before the mapping to machine instructions. Therefore, it is largely unknown whether or not a certain in-lining will have a positive effect. Interestingly, benefits from *WCET*-aware in-lining were rather limited [29]. However, there are also cases where *WCET*-estimation has an impact exceeding expectations. For example, register coloring assigns variables to registers. An abundant amount of algorithms has been proposed for this task. Nevertheless, large improvements are possible with *WCET*-aware register allocation, as can be seen in Fig. 10 [15].

How can such large improvements be explained? Standard register allocation tries to find enough registers so that all variables can be allocated to registers. Typically, this is not feasible and some variables have to be copied to memory (the

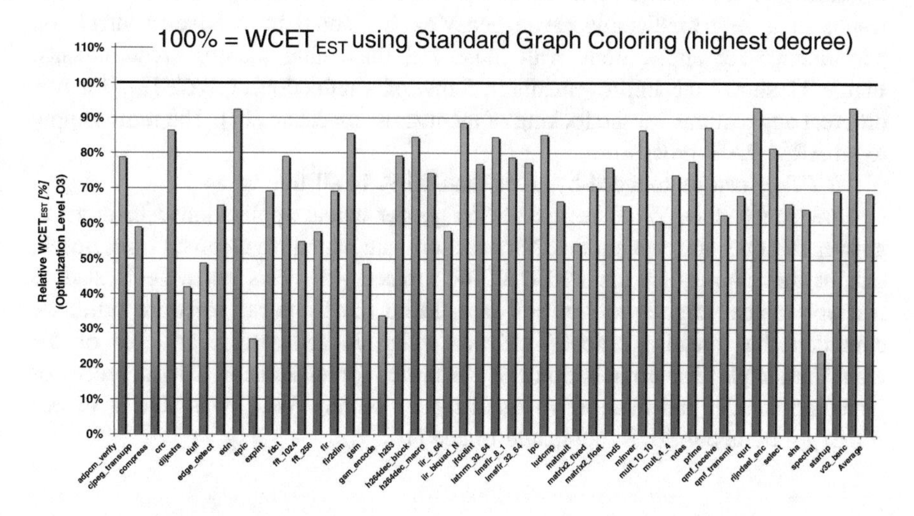

Fig. 10 Result for *WCET*-aware register allocation

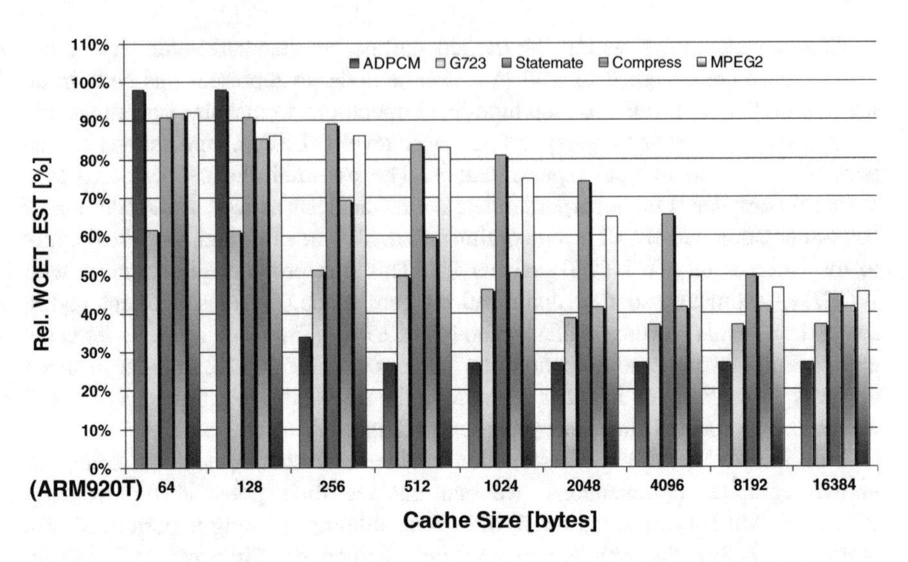

Fig. 11 Result for *WCET*-aware instruction cache locking

so-called register spilling). For a standard compiler, it is difficult if not impossible to predict the effect of spilling a certain variable to memory. For *WCET*-aware register spilling, the effect of selecting a certain variable with respect to the cost function can be predicted. Hence, code quality in terms of this cost function can be improved.

A second example is cache locking. With cache locking, we try to identify objects which are copied into the cache before execution of a certain program is started. Then, the cache is locked. This means that automatic, hardware-controlled replacements are disabled and cache content remains unchanged. Cache locking results in a very predictable cache behavior, but could have adverse effects on performance for applications with large and fluctuating memory requirements. Figure 11 shows the improvements in terms of a reduction of $WCET_{EST}$ for five different applications for the locking of the instruction cache [17]. The results apply to an ARM 920T processor.

$WCET_{EST}$ can be reduced by more than 50 % in all five cases.

How much does WCC reduce $WCET_{EST}$ for actual applications? In order to answer this question, we ran the Democar software which we received from Bosch, one of the partners in our PREDATOR project [45]. This software is said to resemble actual engine control software. From the Democar software suite, we compiled the thread IgnitionSWCSync, which is activated every 90° of the crankshaft angle. The compilation was performed for an Infineon Tricore processor including scratch pad memory (a small tagless buffer). WCC was able to reduce $WCET_{EST}$ by about 50 %, compared to gcc [16].

4.4 Flexible Fault Tolerance

As outlined above, creating reliable hardware/software platforms is one of the challenges in the design of CPS. Traditional fault-tolerance approaches are based on the replication of hardware components [5]. These approaches rely on the addition of redundant processing units, using Dual or Triple Modular Redundancy, and redundant storage using error correcting codes [21] to detect and correct permanent and transient errors occurring during a system's lifetime with a high probability of success.

Typical embedded computers used to control CPS, however, face a number of constraints in time, energy, space, and cost, which render these traditional hardware-based approaches infeasible. Software-implemented fault tolerance, such as SWIFT [46] and redundant multithreading [53], provide lower-cost error detection and correction (EDAC) approaches. Compared to hardware EDAC approaches, however, these software-based approaches mostly rely on redundant execution of code without reserving some of the available hardware units exclusively for redundancy purposes. As a consequence, the computational overhead required to correct a detected error will disturb the expected temporal behavior of a CPS. This could result in deadline misses with potentially far more severe consequences for the overall system compared to the impact of the error which was corrected in the first place.

One of the problems that leads to a high overhead for hardware-implemented as well as early software-implemented fault-tolerance methods is that these approaches neglect to include *semantic information* on the possible worst-case outcome of a specific error. As a consequence, these approaches have to handle all errors alike, which results in the same effort expended to correct fatal errors and errors with no or only negligible effect on a system's operation. A SWIFT approach to overcome some of these efficiency problems can include information about the worst-case severity of a specific error in order to reduce the error correction overhead [23]. A classification of the effect of a given error is employed to determine the mode of error correction to apply. Specifically, this classification enables a SWIFT system to determine *if*, *when*, and *how* to correct an error. This separation of error detection and error correction allows for a number of degrees of freedom to handle errors which do not critically impact system operation, i.e., errors which do not result in a system crash or are externalized to control critical components of a system. These errors may be ignored (*if*), their handling can be delayed (*when*), and, finally, the EDAC system is also able to select among a set of—potentially imperfect—error correction methods (*how*) with different impacts on the timing and energetical overhead of the overall CPS.

Figure 12 shows a possible effect of this separation of error detection and correction by introducing an efficient error classification approach into SWIFT. In the commonly used approach, as shown on top of the figure, error detection is immediately followed by a correction phase, which introduces a significant overhead, e.g., to restore a previous system checkpoint and re-execute calculations or to algorithmically

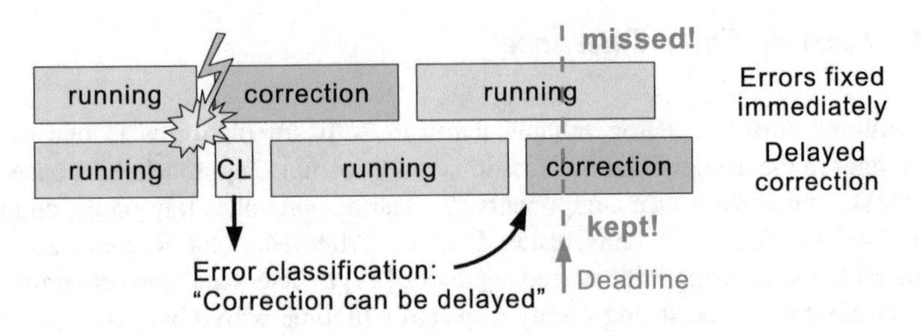

Fig. 12 Temporal behavior of EDAC without and with error classification

correct erroneous data in memory. Due to the introduction of a classification phase, as indicated by the block labeled "CL" in the bottom row of the figure, the handling of non-severe errors can be delayed as shown, or even ignored, which allows the task currently executing on the system to adhere to its real-time constraints.

This classification of possible worst-case error impacts requires information on the *semantics* of a CPS application which is commonly not available using standard software development toolchains. Advanced compiler techniques to support more efficient operation of SWIFT systems, thus, commonly include methods to provide annotations indicating the criticality of an error on the execution of an application. On the one hand, annotations can indicate the criticality of a section of code, as shown for the Relax framework [25] in Fig. 13.

Here, the body of a function is enclosed in a `relax` statement, which indicates that the enclosed code should be handled with relaxed error correction. In case of an error, the correction method applied can vary, as indicated in the `recover` clause of the block. A recovery method of `retry` indicates classical checkpoint-and-restore behavior, whereas the method indicated in Fig. 13 corrects a potential error imprecisely by returning the maximum integer value instead of the expected sum of the passed values.

However, these code-based annotations face a significant disadvantage, since they are unable to capture the *data flow* throughout an application. This results in a large number of annotations that have to be provided by the developer in order to reduce the EDAC overhead and, additionally, is unable to distinguish between different use cases of code, i.e., functions which can handle data with different levels of criticality to the overall application. As a consequence, our FEHLER

```
int sad(int *left, int *right, int len) {
  relax (rate) {
    int sum = 0;
    for (int i = 0; i < len; ++i)
      sum += abs(left[i] - right[i]);
  } recover { return INT_MAX; }
  return sum;
}
```

Fig. 13 Code-based annotations of error criticality and correction approach in Relax (from [25])

```
static inline unreliable int Clip1(int i) {
  unreliable int j;
  if(i<0) j = 0; else if(i>255) j = 255; else j = i;
  return j;
}
```

Fig. 14 Data-centric reliability annotations in FEHLER [13]

project introduced *data annotations*, which can be automatically analyzed and propagated throughout a program's code base by an advanced compiler [13]. The worst-case criticality of an error affecting a data object (a variable or structure) of a program is indicated by an additional type qualifier. As shown in Fig. 14, FEHLER implements two classes of qualifiers.

A variable classified as `reliable` is critical to the execution of a system and demands instantaneous, perfect error correction, whereas a data object classified as `unreliable` can sustain incorrect values without causing critical errors.

One of the major advantages of this data-centric approach is that only a number of key data objects, such as input and output variables, have to be annotated manually. The criticality of related variables in the control and data flow of the program can, in turn, be derived by appropriate static analyses of the program's code. As shown in Fig. 15, the derivation of annotations for non-annotated variables is the result of a data flow analysis, which adheres to a number of safety constraints which prohibit the propagation of possibly erroneous (`unreliable`) values to `reliable` variables which require correct data.

Additional rules require reliable data in key objects such as variables influencing the control flow of a program and pointers.

Many applications of CPS are an ideal target for this data-centric flexible error handling, since they comprise a large amount of signal processing operations. In simple sensors, such as analog-to-digital converters, as well as in complex sensors, such as video cameras and radar systems, inputs coming from the physical environment of a CPS are subjected to noise, i.e., statistical and methodological aberrations of sensor values from the physical value that is measured. As a consequence, the related algorithms and the code derived from these are already tolerant to a given level of imprecision. An analysis of the improvements of the flexible SWIFT implemented in FEHLER in a typical signal processing application, H.264 video decoding, was performed in [49]. Figure 16 shows frames from a video decoded under the influence of transient memory errors.

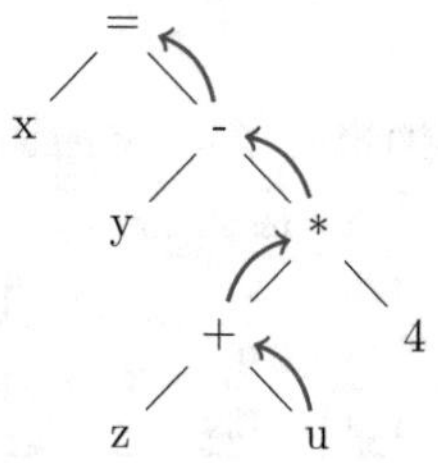

Fig. 15 Safe derivation of type qualifiers in FEHLER

```
unreliable int u, x;
int            y, z;
...
x = y - ( z + u ) * 4
```

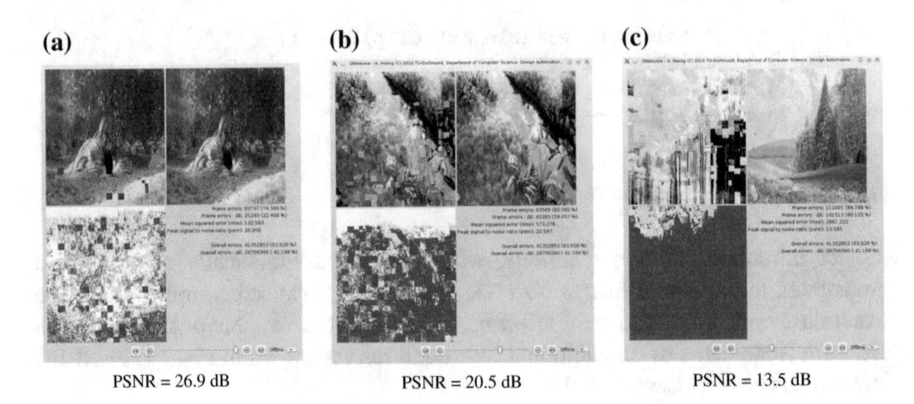

Fig. 16 Comparison of different PSNR values

The amount of error introduced is given as the peak signal-to-noise ratio (PSNR) of the erroneously decoded frame compared to the same frame decoded without errors. The different subfigures of Fig. 16 show the effect of different error rates on the quality of the output. In the example, all errors influencing the decoded picture data have been ignored. As a consequence, the time required for decoding a frame did not change significantly (except for a small error classificiation overhead), resulting in a sustained decoding frame rate. However, depending on the amount of errors introduced, the quality of service (QoS) of the decoder, as indicated by the PSNR rate, decreased significantly from picture (a) to (c). For a CPS, the acceptable QoS of a system under error influence has to be determined by the system designer. Here, additional research is required to ensure the constraining of `unreliable` output values to required safe bounds.

The resource conservation potential of flexible error handling, however, is significant. Table 2 shows the reduction in the amount of main memory that has to be protected by ECC for different video resolutions using FEHLER.

For low resolutions, already 45 % of the data processed by the video decoder could remain unprotected, whereas for a 720p resolution, as much as 63 % of all data memory did not require ECC. In addition to the reduction in protected memory, the timing behavior of the decoder could also be improved. As shown in Fig. 17, the jitter introduced (the number of missed deadlines) by the different approaches to error correction available in FEHLER could be reduced to less than 0.4 % for an error rate of $\lambda = 10^{-16}$ while constraining most jitter to a negligible timing deviation of less than 10^{-5} %.

Table 2 Ratio of reliable to unreliable memory

Video resolution	Memory size of reliable data	Memory size of unreliable data
176 × 144	92,908 bytes (55 %)	76,184 bytes (45 %)
352 × 288	228,784 bytes (43 %)	304,280 bytes (57 %)
1280 × 720	1,623,104 bytes (37 %)	2,764,952 bytes (63 %)

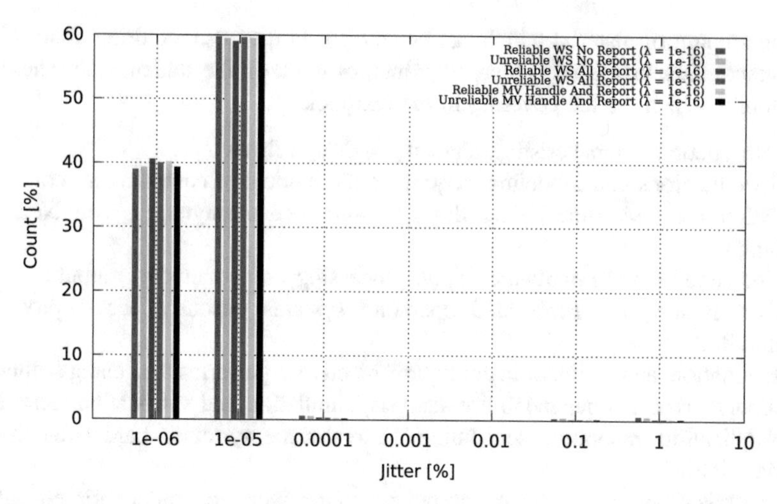

Fig. 17 Reduction of decoding jitter

Recent approaches do not only concentrate on efficient error handling; instead, they deliberately accept errors in computations in order to provide a more efficient execution of software. These *approximate computing* approaches allow for the reduction of energy and execution time by enabling the near- or sub-threshold operation of semiconductors, which can result in timing violations in clocked circuits as well as value aberrations in memory cells. Techniques such as Probabilistic CMOS (PCMOS, [9]) implement hardware which exploits these characteristics, e.g., probabilistic adders and multipliers. Based on these components, software based approaches such as EnerJ [48] and an extended version of the FEHLER type qualifiers [24] can be used to determine operations which can accept imprecise results and, thus, can be safely executed using probabilistic hardware components. While approximate computing techniques are a comparatively new area of research, recent publications from physical atmosphere and climate modelling [11] show a promising potential for significant reductions of energy and runtime overhead.

4.5 Education

We mentioned CPS education as a crucial area.

We have provided, for many years in a row, embedded systems as a subset of the scope of CPS as a concentration (set of electives) of our computer science program. Our course on embedded systems plays a key role in this concentration. For this course, we have carefully selected core topics of embedded systems. These core topics have traditionally been described in disjoint locations and have been difficult to access for students. We have therefore written a textbook [31] which comprises the topics which we consider to be the most relevant in this area, limiting ourselves to the amount of material which can be covered in a course requiring about 135 h (corresponding to 9 European credit points) of work on the student side. These are the topics which we are covering in the textbook:

1. Introduction: characteristics, challenges, design flows
2. Specifications and modeling: requirements, models of computation, communication state machines, data flow, discrete event languages, von Neumann languages
3. Embedded system hardware: input, processing units, memory, output
4. System software: Embedded operating systems, resource access protocols, middleware
5. Evaluation and validation: multiple objectives, performance, energy, thermal models, risk and dependability analysis, simulation and verification techniques
6. Application mapping: scheduling in real-time systems, hardware-software partitioning
7. Optimization: task level concurrency management, compilers for embedded systems, power and thermal management
8. Test: test procedures, evaluation of test pattern sets, design for testability

Using the textbook, we ran the embedded system course for more than a decade and finally decided to record the lectures. The lectures are now available online at youtube [32].

As a result of the availability of the videos, students started to suffer from a very serious case of procrastination: they skipped the lectures, assuming that they could learn from the videos just before the finals. Of course, watching around 40 h of technical videos just before the finals and still understanding the material does not work. Only few students attended the finals. In order to cope with this problem, we adopted the flipped classroom concept: we requested the students to watch the videos at home and to work on worksheets during class meetings [33]. Using these worksheets, we trained the teamwork capabilities and helped them to understand the material at a level that allows them to pass the finals.

Teaching embedded systems as a concentration of a computer science program has to cope with tight constraints on the workload. If we want to go further and include more CPS topics, we have to carefully evaluate advantages and limitations of using a concentration of an established program versus creating a new one on cyber-physical systems. Table 3 lists advantages and limitations of both approaches. Obviously, the usual credit point and program constraints have to be met. In Europe, Bologna regulations apply to most universities.

The bottom line is that only a program exclusively focusing on CPS has enough headroom to cover CPS core content. Let us consider an example of a curriculum design of such an integrated program. We have opted for a program at both the undergraduate as well as the graduate level, since we believe that a CPS master program would not allow the workload necessary to cover key areas of CPS. Even for a program which includes undergraduate as well as graduate education, we observed a tight constraint on the workload, which made it necessary to select between possible concentrations. The current proposal focuses on an integration of

Table 3 Tradeoffs between a separate CPS program and CPS as a concentration of an available program

	Concentration	Separate CPS-program
Well-known degrees	+	–
Enough headroom for teaching integrated CPS/ES material	–	+
Headroom for more depth in physics, mechanical engineering, biology	–	+
Effort for introduction	Small	Large
No. of faculty members required	Moderate	Larger
Building community?	Difficult	Easier
Feasible at all?	For ES ok, for CPS questionable	Yes, but there are still tight workload constraints

computer science and communication aspects of electrical engineering. The following list contains the subjects which we have included in our draft curriculum for a 4.5 year integrated CPS program.

- Undergraduate level
  - Computer science (31.11 %): data structures, algorithms, programming, computer structures, operating systems, networks and distributed systems, embedded systems, information systems
  - Electrical engineering (20 %): Foundations of electrical engineering, system theory, communication networks, theory of communication
  - Mathematics (17.22 %): engineering mathematics, statistics
  - Logistics (5 %): Flow of materials, automation
  - Integrated CPS content (16.67 %): project, integrated CPS course, Bachelor thesis
  - Electives/concentration (10 %): data mining, mechatronics, data analysis, signal processing, smart grid, …
- Graduate level
  - Computer science (21.67 %): theoretical computer science, modeling, simulation
  - Electrical engineering (15 %): advanced communication modeling
  - Logistics (6.67 %): advanced automation
  - Integrated CPS content (33.33 %): seminar, project, Master thesis
  - Electives/concentration (16.67 %): intelligent systems, robotics, advanced network concepts
  - Non-technical subjects (6.66 %): legal and economical aspects of CPS

However, such a new program requires a larger effort, compared to CPS as a concentration in a computer science or electrical engineering program. We have proposed such a new program to the head of our University. At the time of writing this text, acceptance is pending.

5 Conclusion

In this chapter, we have shown that the term Cyber-Physical Systems refers to a large percentage, if not the majority, of future applications of ICT. A huge amount of industrial sectors is affected by the availability of CPS. However, quite a number of challenges have to be overcome, which means that an extremely large research and development effort is required. In order not to waste resources, it is necessary to work against the possible fragmentation of research efforts and to highlight commonalities. In the final main part of this chapter, we have shown how some of the issues can be addressed and how the challenges can at least be partially overcome.

Acknowledgement Some of the work described in this chapter has been supported by Deutsche Forschungsgemeinschaft in the context of our Collaborative Research Center SFB 876 and the Priority Program SPP 1500 on Dependable Embedded Systems. We do also acknowledge the support of the European Community through the ArtistDesign network of excellence and the PREDATOR project under grant agreement no. 216008.

References

1. Absint: aiT worst-case execution time analyzers (2015). http://www.absint.de/ait
2. Agogino, A.: Distributed coordination of agents for air traffic flow management (2009). http://cps-vo.org/node/954
3. Akatech: Cyber-physical systems. Driving Force for innovation in mobility, health, energy and production (2011). http://www.acatech.de/de/publikationen/stellungnahmen/kooperationen/detail/artikel/cyber-physical-systems-innovationsmotor-fuer-mobilitaet-gesundheit-energie-und-produktion.html
4. Auffermann, C., Kamagaev, A., Nettsträter, A., ten Hompel, M., Vastag, A., Verbeek, K., Wolf, O.: Cyber physical systems in logistics. http://www.effizienzcluster.de/files/9/5/938_scientific_paper_cyber_physical_systems_in_logistics.pdf
5. Aviziens, A.: Fault-tolerant systems. IEEE Trans. Comput. **C-25**(12), 1304–1312 (1976). doi:10.1109/TC.1976.1674598
6. Brisk, P.: System support for generally programmable digital microfluidic biochip devices (2011). http://cps-vo.org/node/1029
7. Bullo, F., et al.: Dynamic routing and robotic coordination for oceanographic adaptive sampling (2013). http://cps-vo.org/node/9357
8. Cavusoglu, M.C.: A framework for validation and monitoring of robotic surgery systems (2010). http://cps-vo.org/node/1028
9. Chakrapani, L., Muntimadugu, K., Lingamneni, A., George, J., Palem, K.: Highly energy and performance efficient embedded computing through approximately correct arithmetic. In: Proceedings of CASES, pp. 187–196. ACM (2008)
10. Cousot, P.: Abstract Interpretation. ACM Comput. Surv. **28**(2), 324–328 (1996). doi:10.1145/234528.234740. http://doi.acm.org/10.1145/234528.234740
11. Düben, P.D., Joven, J., Lingamneni, A., McNamara, H., De Micheli, G., Palem, K.V., Palmer, T.N.: On the use of inexact, pruned hardware in atmospheric modelling. Philos. Trans. Royal Soc. London A: Math. Phys. Eng. Sci. **372**(2018) (2014). doi:10.1098/rsta.2013.0276
12. Dunstan, W.: Computer and GPS navigation system for an autonomous ocean vessel. In: Proceedings of the Third Australian and New Zealand Conference on Intelligent Information Systems, ANZIIS-95, pp. 316–323 (1995). doi:10.1109/ANZIIS.1995.705758

13. Engel, M., Schmoll, F., Heinig, A., Marwedel, P.: Unreliable yet useful—reliability annotations for data in cyber-physical systems. In: Proceedings of the 2011 Workshop on Software Language Engineering for Cyber-physical Systems (WS4C). Berlin, Germany (2011)
14. European Commission: ICT 2014—Information and Communications Technologies, call H2020-ICT-2014-1 (2014). http://ec.europa.eu/research/participants/portal/desktop/en/opportunities/h2020/topics/78-ict-01-2014.html
15. Falk, H.: WCET-Aware Register Allocation Based on Graph Coloring. In: The 46th Design Automation Conference (DAC), pp. 726–731. San Francisco, USA (2009)
16. Falk, H., Kleinsorge, J.C.: Optimal static WCET-aware scratchpad allocation of program code. In: The 46th Design Automation Conference (DAC), pp. 732–737. San Francisco (2009)
17. Falk, H., Plazar, S., Theiling, H.: Compile time decided instruction cache locking using worst-case execution paths. In: International Conference on Hardware/Software Codesign and System Synthesis (CODES + ISSS), pp. 143–148. Salzburg, Austria (2007)
18. Falliere, N., Murchu, L.O., Chien, E.: W32. stuxnet dossier. White paper, Symantec Corp., Security Response (2011)
19. Federal Goverment of Germany: The new high-tech strategy—innovations for Germany (2014). http://www.bmbf.de/pub/HTS_Broschuere_engl_bf.pdf
20. Fraunhofer Institute for Material Flow and Logistics: InBin—Intelligent Bin (2015). http://www.iml.fraunhofer.de/en/fields_of_activity/automation_embedded_systems/Products/InBin.html
21. Hamming, R.W.: Error detecting and error correcting codes. Bell Syst. Tech. J. **29**(2), 147–160 (1950). doi:10.1002/j.1538-7305.1950.tb00463.x
22. Hardkernel: Odroid-XU3 (2014). http://www.hardkernel.com/main/products/prdt_info.php?g_code=G140448267127
23. Heinig, A., Engel, M., Schmoll, F., Marwedel, P.: Using application knowledge to improve embedded systems dependability. In: Proceedings of the Workshop on Hot Topics in System Dependability (HotDep 2010). USENIX Association, Vancouver, Canada (2010)
24. Heinig, A., Mooney, V.J., Schmoll, F., Marwedel, P., Palem, K., Engel, M.: Classification-based improvement of application robustness and quality of service in probabilistic computer systems. In: Proceedings of ARCS 2012—International Conference on Architecture of Computing Systems. Munich, Germany (2012)
25. de Kruijf, M., Nomura, S., Sankaralingam, K.: Relax: An architectural framework for software recovery of hardware faults. In: Proceedings of the 37th Annual International Symposium on Computer Architecture, ISCA '10, pp. 497–508. ACM, New York, NY, USA (2010). doi:10.1145/1815961.1816026
26. Lee, E.A.: The future of embedded software. ARTEMIS Conference, Graz (2006). http://ptolemy.eecs.berkeley.edu/presentations/06/FutureOfEmbeddedSoftware_Lee_Graz.ppt
27. Lee, E.A.: Computing foundations and practice for cyber-physical systems: a preliminary report. Technical Report UCB/EECS-2007-72, EECS Department, University of California, Berkeley (2007). http://www.eecs.berkeley.edu/Pubs/TechRpts/2007/EECS-2007-72.html
28. Libuschewski, P., Siedhoff, D., Timm, C., Gelenberg, A., Weichert, F.: Fuzzy-enhanced, real-time capable detection of biological viruses using a portable biosensor. In: Proceedings of the International Joint Conference on Biomedical Engineering Systems and Technologies (BIOSIGNALS), pp. 169–174 (2013)
29. Lokuciejewski, P., Gedikli, F., Marwedel, P., Morik, K.: Automatic WCET reduction by machine learning based heuristics for function inlining. In: 3rd Workshop on Statistical and Machine Learning Approaches to Architectures and Compilation (SMART), pp. 1–15 (2009)
30. Marwedel, P.: Embedded System Design. Kluwer Academic Publishers, Berlin (2003)
31. Marwedel, P.: Embedded System Design—Embedded Systems Foundations of Cyber-Physical Systems. Springer, Berlin (2011)
32. Marwedel, P., Engel, M.: Videos of the course cyber-physical system fundamentals at TU Dortmund (2012). http://www.youtube.com/user/cyphysystems
33. Marwedel, P., Engel, M.: Flipped classroom teaching for a cyber-physical system course—an adequate presence-based learning approach in the internet age. In: Proceedings of the Tenth European Workshop on Microelectronics Education (EWME). IEEE, Tallinn, Estonia (2014)

34. Morik, K., et al.: Collaborative research center on resource constrained machine learning (2015). http://www.sfb876.tu-dortmund.de
35. Mosterman, P.J.: Hybrid dynamic systems: modeling and execution. In: Fishwick, P.A. (ed.) Handbook of Dynamic System Modeling, CRC Press (2007)
36. National Instruments: Overview of structural health monitoring solutions (2014). www.ni.com/white-paper/8426/en/pdf
37. National Science Foundation: Cyber-Physical Systems (CPS) (2010). http://www.nsf.gov/pubs/2010/nsf10515/nsf10515.htm
38. Neugebauer, O., Libuschewski, P., Engel, M., Mueller, H., Marwedel, P.: Plasmon-based virus detection on heterogeneous embedded systems. In: Proceedings of Workshop on Software and Compilers for Embedded Systems (SCOPES) (2015)
39. Okuda, R., Kajiwara, Y., Terashima, K.: A survey of technical trend of ADAS and autonomous driving. In: Proceedings of Technical Program—2014 International Symposium on VLSI Technology, Systems and Application (VLSI-TSA), pp. 1–4 (2014). doi:10.1109/VLSI-TSA.2014.6839646
40. Otter, M., Winkler, D.: Modelica Overview (2013). https://www.modelica.org/education/educational-material/lecture-material/english/ModelicaOverview.ppt
41. Pappas, G.: Science of cyber-physical systems bridging CS and control (2012). http://cps-vo.org/node/5876
42. Penemon Institute: 2014 Global report on the cost of cyber crime (2014). http://www.ponemon.org/library/2014-global-report-on-the-cost-of-cyber-crime
43. Peter Greenhalgh, ARM: Big.LITTLE processing with ARM Cortex-A15 & Cortex-A7 (2013). http://www.arm.com/files/downloads/big_LITTLE_Final_Final.pdf
44. Potop-Butucaru, D., de Simone, R., Talpin, J.P.: The synchronous hypothesis and synchronous languages. In: Richard, Z. (ed.): Embedded Systems Handbook, CRC Press (2006)
45. PREDATOR Consortium: Predator—design for predictability and efficiency. http://www.predator-project.eu/
46. Reis, G.A., Chang, J., Vachharajani, N., Rangan, R., August, D.I.: Swift: software implemented fault tolerance. In: Proceedings of the International Symposium on Code Generation and Optimization, CGO '05, pp. 243–254. IEEE Computer Society, Washington, DC, USA (2005)
47. Ross, P.E.: When will software have the right stuff? Spectr. IEEE **48**(12), 38–43 (2011). doi:10.1109/MSPEC.2011.6085781
48. Sampson, A., Dietl, W., Fortuna, E., Gnanapragasam, D., Ceze, L., Grossman, D.: EnerJ: approximate data types for safe and general low-power computation. In: Proceedings of PLDI, pp. 164–174. ACM, New York, NY, USA (2011)
49. Schmoll, F., Heinig, A., Marwedel, P., Engel, M.: Improving the fault resilience of an H.264 decoder using static analysis methods. ACM Trans. Embed. Comput. Syst. (TECS) **13**(1s), 31.1–31.27 (2013)
50. Tabuada, P.: Is it about time for control? 2012 NSF CPS PI meeting, http://cps-vo.org/node/5997 (2012)
51. Taha, W., Cartwright, R.: Some Challenges for Model-Based Simulation. The 4th Analytic Virtual Integration of Cyber-Physical Systems Workshop, Vancouver (2013)
52. TriQuint Semiconductor Inc.: FAQ 11: what is the MTBF for gallium arsenide devices? (2010). http://www.triquint.com/company/quality/faqs/faq_11.cfm
53. Vijaykumar, T.N., Pomeranz, I., Cheng, K.: Transient-fault recovery using simultaneous multithreading. In: Proceedings of the 29th Annual International Symposium on Computer Architecture, ISCA '02, pp. 87–98. IEEE Computer Society, Washington, DC, USA (2002). http://dl.acm.org/citation.cfm?id=545215.545226
54. Weiser, M.: Ubiquitous computing (2003). http://www.ubiq.com/hypertext/weiser/UbiHome.html

Cyber-Physical Objects as Key Elements for a Smart Cyber-City

Riccardo Petrolo, Valeria Loscri and Nathalie Mitton

Abstract The continuous growth of the urban population has generated a drastic expansion of our cities. Nowadays, indeed, more than 50 % of the world's population is urban, and they forecast that it will reach 70 % by 2050. Therefore, cities need to be ready to accommodate this huge amount of citizens and to face new challenges (e.g., traffic congestion, air pollution, waste management, etc.). The concept of cyber-physical systems, as integration of computation and physical processes, can help toward the realization of real smart cities capable to ensure sustainability and efficiency. To this purpose, this chapter investigates the cyber-physical system (CPS) and their cyber-physical object (CPO) as key units, in the context of a smart city concept. We survey the smart city vision, providing information on the main requirements, the open challenges, and highlighting the benefits; we also browse the European Commission initiatives for smart cities and some pilot projects that are in development.

Keywords Smart city · Cyber-physical systems · Cyber-physical objects

1 Introduction

The continuous growth of cities started with the urbanization phenomena in the late eighteenth century. Since that time, more and more people moved from rural to urban areas in order to access major opportunities for jobs, education, housing, and transportation.

R. Petrolo (✉) · V. Loscri · N. Mitton
Parc scientifique de la Haute Borne, Inria Lille-Nord Europe,
Villeneuve d'Ascq, France
e-mail: riccardo.petrolo@inria.fr

V. Loscri
e-mail: valeria.loscri@inria.fr

N. Mitton
e-mail: nathalie.mitton@inria.fr

A. Guerrieri et al. (eds.), *Management of Cyber Physical Objects in the Future Internet of Things*, Internet of Things,
DOI 10.1007/978-3-319-26869-9_2

Following this trend, in 2010, 50 % of the world's population was living in cities [1]; as shown in Fig. 1, in the developed world the proportion of people living in cities is higher than 75 %. In China, even though the proportion of people living in cities is under 50 %, the total number of urban dwellers is greatest (559 million).

Fast Co. Design [2] forecasts that in 2050, 70 % of the world's population will be urban; therefore, it is crucial, for cities, to be ready to accommodate this huge amount of citizens and to face new challenges (e.g., traffic congestion, air pollution, waste management, water monitoring, and so on).

In this context, information and communication technologies (ICT), together with local governments and private companies, play a key role for implementing innovative solutions, services, and applications to make the *smart city* a reality [3].

Hancke et al. [4] define a smart city as a *city which functions in a sustainable and intelligent way, by integrating all its infrastructures and services into a cohesive whole and using intelligent devices for monitoring and control, to ensure sustainability and efficiency.*

This definition makes evident that the smart city concept (Fig. 2) needs interoperability between the different application domains, that are, nowadays, mainly closed and vertically integrated (e.g., smart mobility, smart living, etc.).

Aloi et al. [5] outlined the requirements for a smart city can be classified into two different types:

1. *Service/application*, considered from the point of view of the citizens.
2. *Operational*, seen from the city authorities and network administrators viewpoint.

Concerning the *service/application* aspects, the end-user devices equipped with multiple radio technologies and several sensors and actuators deployed all over the cities, make possible the individuation of novel services and applications for the

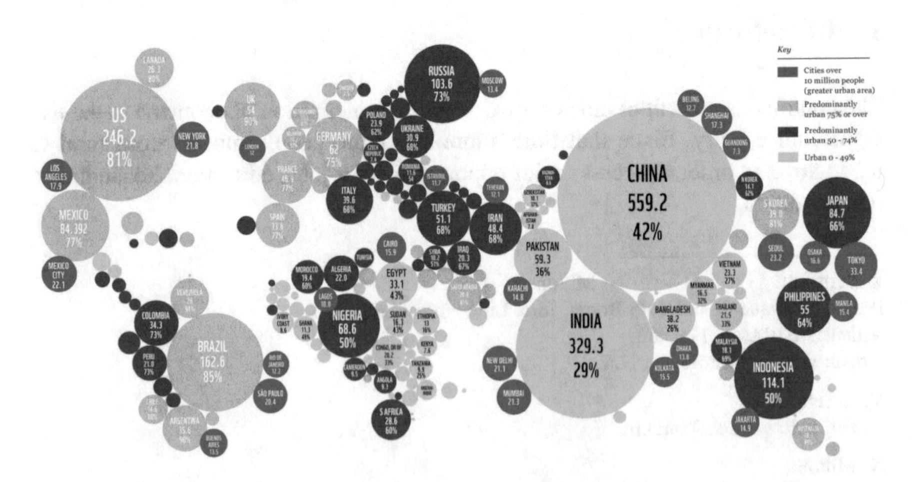

Fig. 1 Number of people together with percentage of population living in cities in each country in 2010 [41]

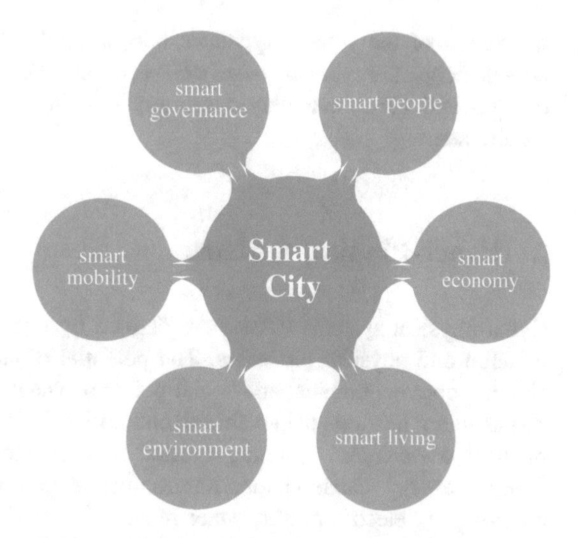

Fig. 2 Smart city concept

citizens. These services will have specific features, like: (a) *user-centric*: based on the specific context and the preferences of the users, (b) *ubiquitous*: reachable everywhere and from any devices, and (c) *highly integrated*: based on the integration of services and data from several and different applications or on the social cooperation of multiple users. Of course, beyond the citizens, also the stakeholders of a city, like educational institutions, health-care and public safety providers, governmental organizations, etc., will be in conditions to exploit the key features of these new services that make the city more sustainable.

On the other hand, the smart city concept considered from the point of view of the administrations and the network providers are translated into a network infrastructure, i.e.,: (a) *highly interconnected*: by overcoming the heterogeneity of the devices and the Internet of Things (IoT) platforms, it is possible to provide ubiquitous connectivity, (b) *cost-efficient*: the deployment and organization of the network should be as much automatic as possible and independent from the human intervention, (c) *energy-efficient*, able to realize an efficient resource utilization, in order to meet the main requirements of *green* applications, and (d) *reliable*: that connectivity, the ubiquity of the network should be guaranteed above all in the case of exceptional and adverse conditions. The real scenario we can observe at the moment is characterized with a high level of *fragmentation* of technologies, lack of ubiquity in terms of both connectivity and coverage, due to the plethora of technologies and devices present in a city. This *fragmentation* is mainly due to the presence of many access networks usually managed by different operators (i.e., Universal Mobile Telecommunications System—UMTS, Worldwide Interoperability for Microwave Access—WiMAX, WiFi, etc.).

The requirements aforestated trace the need of systems highly interconnected, able to communicate with the surrounded environment in order to learn, self-organize and react. To the follow, we will introduce what we envisioned as a potential "horizontal-integrator" enabler that can play a crucial role for the realization of the smart city: the *cyber-physical system (CPS)* concept and their key-elements, the *cyber-physical object (CPO)*. We will analyze as the key features of *CPS* and *CPO*, can play a primary role in the fulfillment of the smart city requirements.

2 Cyber-Physical Systems

Cyber-physical systems (CPSs) are defined by Lee [6] as the integration of computation and physical processes. The potential of such systems is enormous considering both economical and social points of view; just thinking, for example, the disparate applications achievable thanks to CPSs, to name a few, high confidence medical devices and systems, assisted living, traffic control and safety, process control, energy conservation, environmental control, and critical infrastructure control (e.g., electric power, water resources, and communication systems).

Wu et al. [7] discuss the unique features of CPS introducing also some technical challenges:

- *Cross-domain and cross-network*. Multiple types of sensors will be adopted at the same time; these cross-domain data will be exchanged over heterogeneous networks.
- *Embedded and mobile sensing*. Sensors are no longer static and may have high-degree mobility through carriers such as smartphones and vehicles, introducing then uncertainty due to the variability of the sensing coverage. Intelligent discovery mechanisms are required to analyze these mobile data.
- *Elastic load*. With the maturity of cloud computing, the *pay-as-you-go* concept, introduced by Banerjee et al. [8], is likely to be adopted in CPS to serve storage, computing, and communication needs. This allows CPS developers to focus on their own work and users to choose the part of CPS applications that they really want.
- *Accumulated intelligence*. Data in CPS may have high dynamics and uncertainty, therefore learning and data mining technologies can be useful to retrieve *knowledge*.
- *Interactions among many objects*. A lot of sensor–sensor, sensor–user, sensor–actuator, user–user, and user–actuator interactions may occur in CPS applications, therefore a flexible communication channel, like the Internet, is required.

Figure 3 shows the differences between two different approaches, middleware and CPS. The first solutions (Fig. 3a) follow a vertical architecture, indeed servers, at the virtual layer, use the network to analyze data gathered from the sensors, and then send command to the actuators. On the other hand, the CPS solution (Fig. 3b), is more independent, devices are all the same level horizontally integrated around different networks.

The above features manifest the differences between CPSs and traditional systems (e.g., desktop computer, wireless sensor networks (WSNs), etc.) and they also open

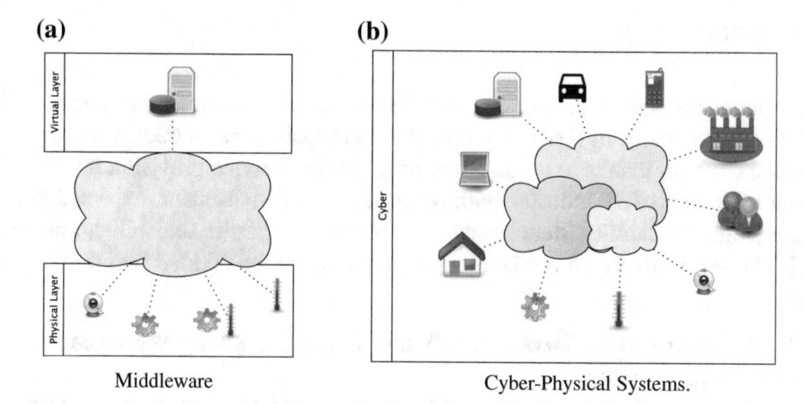

Fig. 3 Middleware versus CPS approach

new research directions on the field. It is reasonable that *networking* and *computing* aspects play a key role within the CPSs evolution; below we review the technical challenges and progresses that have been made with regards to these issues.

2.1 Networking

As own enabler of the communication between the different actors of the CPSs, the network aspect plays a key role. In the past years, researchers have focused on wireless communication and network, making significant progress in the fields of *mobile ad hoc network* (MANET) and *wireless sensor network* (WSN). CPS would take advantage from expertise in the areas of MANET and WSN since they are quite similar in many networking aspects even if there are some major differences. Roughly, while MANET is for ad hoc communications [9] and WSN is designed for delivering sensor data [10], CPS aims to construct intelligence across different domains like sensing data, crosses multiple sensor network and the Internet.

Therefore a CPS has to combine WSNs with the Internet, and a lot of inter-working issues have to be resolved. Indeed, ubiquitously deployed Internet protocols such as HTTP, TCP, or even IP are too complex and resource-demanding [11]. The *μIP* [12] includes a low-power link built on IEEE 802.15.4 for small embedded devices. IETF introduces 6LoWPAN [13] which defines mechanisms capable to fragment and to compress the header of IPv6 datagrams. In [14] authors consider multiple WSNs connected by IPv6-based *border routers* through IP link, including Ethernet, WIFi, GPRS, and satellites. COAP [15] is an application layer protocol designed for energy constrained devices. It deals with constrained RESTful environments, providing a lightweight alternative to HTTP.

All the technologies above presented, open an opportunity for future CPSs, since cross-domain end-to-end communication among *objects* is possible.

2.2 Computing

We believe that cloud computing techniques can play a primary role toward the CPS's development. In the past years, cloud computing has attracted the attention across the world, thanks to its ability of transforming service provision models over the entirely current IT industry with reduced upfront investment, expected performance, high availability, tremendous fault tolerance, infinite scalability, and so on [16]. The services offered by the cloud computing can be divided into three layers: [17]:

- *Infrastructure as a Service* (IaaS) that offers computing resources such as processing or storage.
- *Platform as a Service* (PaaS) that offers particular platform to software developers according to their specification.
- *Software as a Service* (SaaS) that offers software applications to be accessed and used by end-users.

This service *isolation* enables autoscaling and automanaging capabilities that are crucial for the future CPSs.

In addition to the above main layers, some others are also introduced and discussed in the literature such as *Data as a Service* (DaaS) [18], *Network as a Service* (NaaS) [19], and *Identity and Policy Management as a Service* (IPMaaS). The *Everything as a Service* model (XaaS) [8] promotes the "*pay-as-you-go*" method, allowing the consumption of a service by paying only for the amount of resources used. This concept is also at the base of the *Sensing as a Service* model, introduced by Perera et al. [20], in which authors delineate four conceptual layers:

- *Sensor and Sensor Owners Layer*: to manage sensors and their possible publication into the cloud.
- *Sensor Publishers* to detect available sensors and get permission to publish them into the cloud.
- *Extended Service Providers* to select sensors from multiple publishers based on customer's requirements.
- *Sensor Data Consumers Service Providers* that need to register to consume sensors data.

The advantages and benefits promised by the *SeaS* model are numerous, to name a few: *sharing and reusing sensor data* (no need to deploy other sensors, it is possible to access sensors already deployed by paying a fee to the owner), *reduction of data acquisition cost* due to the shared nature, *collect data previously unavailable* (companies are stimulated to "sell" data).

The inherent features that the SeaS paradigm is able to offer, perfectly match with the smart city requirements. First, the possibility to reuse devices and resources already available; cities are currently disseminated with many sensors and "communicating" devices that may be effectively and properly "re-used." This will embrace the philosophy expressed very well by Jochen Kreusel "Many of the building blocks for creating smart cities are already available. It is an ongoing evolution rather than a disruptive change."[1] Second, the "data shared" nature, intrinsic in SeaS, that would face in some way the problem of the big data intrinsically correlated with an ICT City.

The SeaS together with the other models of cloud computing, represent a good approach to be adopted in CPSs to serve storage, computing, and communication needs.

2.3 Actors of CPSs: Cyber-Physical Objects

In this section, we will try to "dissect" the CPS in order to individuate the atomic units of the CPS. In the following, we will refer to these elementary units/objects that constitute an important block of the CPSs as *cyber-physical object (CPO)*. In order to be integrated in sophisticated and ubiquitous CPSs, *objects* have to self-organize, learn, and react; they will be accessible via the network and queried in order to facilitate everyday life, at home, in the office, and in leisure. Examples of objects, to name some, can range from monitoring devices and sensors, home appliances, lifts, cars, buses, traffic light, parking meters, containers, security cameras, locks, alarms, water valves, wind turbines, drills, retail objects, and retail selves [21].

In [22], Kortuem et al. differentiate smart objects following three design dimensions:

- *awareness* smart object's ability to understand events and human activities occurring in the physical world.
- *representation* refers to a smart object's application and programming model.
- *interaction* denotes the object's ability to converse with the user in terms of input, output, control, and feedback.

The combination of the Internet and emerging technologies as near-field communications, real-time localization, and embedded sensors let us transform everyday objects into *smart objects* that can understand, communicate, interact, and react to their environment. Such objects are building blocks for the CPSs and enable novel computing applications.

In contrast to RFID technology—main actors of the *Internet of Things* [23] success—smart objects carry chunk of application logic that let them make sense of their local situation and interact with human users. They sense, log, and interpret what is occurring within themselves and the world, act on their own, intercommunicate with each other, and exchange information with people.

[1]http://www.abb-conversations.com/2014/10/smart-cities-intelligent-solutions-for-future-generations.

The goal of the Internet of Things is to enable things to be connected *anytime*, *anyplace*, with *anything* and *anyone* ideally using any path/network and any service.

As stated [24] the Internet of Things is a new revolution of the Internet. Objects make themselves recognizable and they obtain intelligence by making or enabling context-related decisions thanks to the fact that they can communicate information about themselves. They can access information that has been aggregated by other things, or they can be components of complex services.

Different technologies are available to connect an object, already well spread in our everyday life environment like RFID in subway, 3G with our phone, WiFi for Internet at home, etc. However, due to the very heterogeneous landscape in terms of hardware capabilities/constrains and network protocols, IP-based access is required as unifying network layer in order to turn smart objects into *Internet-connected objects* (ICOs).

Hauswirth et al. [25] point out on the need of a *semantic representation* in order to understand the data which comes out and goes into the ICO interfaces. This "*data exchange layer*" may influence discovery and routing approaches and it will be crucial to enable scalability from an application's point of view as nobody will be able to deal with the number of ICOs efficiently and scalable without such layer. The necessary technologies are already being developed and deployed: Linked Data [26] and the Resource Description Format (RDF) [27] are accepted standards in the web and provide a general model. However, these technologies need to be condensed into lighter forms in order to be used on resource-constrained devices, very much in the same spirit as CoAP was done for the service side. With light-weight semantics, ICOs will be the first-class citizens of an interned-wide semantic database that can easily be indexed, searched, and used using standard web technologies.

In order to manage those ICOs, several IoT platforms have been introduced in the literature; we review some of the most representative without pretending to be exhaustive:

- *GSN*[2] (Global Sensor Networks), provides a flexible java middleware to address the challenges of sensor data integration and distributed query processing. It lists all the available sensors in a combo box, which users need to select. GSN's purpose is to make applications hardware-independent and the changes and variations invisible to the application. Its main limitation is a lack in metadata semantics.
- *LSM*[3] (Linked Sensor Middleware) is a platform that bridges the live real-world sensed data and the semantic web thanks to many functionalities such as, wrappers for real-time data collection and publishing; data annotation and visualization; and a SPARQL endpoint for querying unified linked stream data and linked data. However, it does not offer tools for manipulating data.

[2]https://github.com/LSIR/gsn.

[3]http://lsm.deri.ie.

- Fortino et al. [28] propose a multilayered agent-based architecture for the development of proactive, cooperating, and context-aware smart objects. This architecture takes into account a wide variety of smart objects, from reactive to proactive, from small to very large, from stand-alone to social.
- *Sensor-Cloud* [29] aims at managing physical sensors by connecting them to the cloud, providing the service instances (virtual sensors) in an automatic way in the same fashion as these virtual sensors are effectively part of the IT resources. The generation of the services implies that the sensor devices and service templates (used to create the virtual sensors) and metadata should be first described by using SensorML.
- *OpenIoT*,[4] a joint effort of several contributors to IoT-based applications according to a cloud computing delivery model, provides a cloud-based middleware infrastructure to deliver on-demand access to IoT services issued from multiple platforms. It can opportunely collect and filter data from the Internet-connected objects. Its main limitation is the lack of an ontology to describe smart city concepts.
- *Xively*[5] (formerly Cosm and Pachube) offers a public cloud that simplifies and accelerates the creation, deployment, and management of sensor in scalable way. Its main constraint is due to the limitation to manage and to retrieve data just from own devices.

These approaches strengthen the importance of the cloud solutions, but a lot of new enhancements are still needed to realize platforms ready to manage a smart city. To fill this gap, in the last years the European Commission, funded several projects; some of the most representative are described in the section below.

2.4 European Commission Initiatives for Smart Cities

The enormous interest that the smart city concept has acquired is witnessed from the several initiatives that also the European Commission has activated.

The European Innovation Partnership on *Smart Cities and Communities—(EIP-SCC)* focuses on the integration of industry, citizen, and cities to try to improve the sustainability of the urban life through integrated solutions. The Seventh Framework Programme for Research and Technological Development of the European Commission funded different projects under the call smart city, in order to correctly identify and address smart city issues and challenges. Without pretending to be exhaustive, we will present some of the most representative projects that have been proposed in the European FP7 calls. *ClouT*[6] uses cloud computing as

[4]http://openiot.eu.

[5]http://xively.com.

[6]http://clout-project.eu.

an enabler to bridge the Internet of Things with the **Internet of People** via the *Internet of Services*, to establish an efficient communication and collaboration platform exploiting all possible information sources to make the cities smarter and to help them facing the emerging challenges such as efficient energy management, economic growth, and development. ClouT will provide infrastructures, services, tools, and applications that will be reused by different city stakeholders such as municipalities, citizens, service developers, and application integrator, in order to create, deploy, and manage user-centric applications taking benefit of the latest advances in internet of things and cloud domains.

SOCIOTAL[7] aims to design and provide key enablers for a reliable, secure, and trusted IoT environment that will enable creation of a socially aware citizen-centric Internet of Things by encouraging people to contribute their IoT devices and information flows. It will provide the technosocial foundations to unlock billions of new IoT information streams taking a citizen-centric IoT approach toward creation of large-scale IoT solutions of interest to the society. By equipping communities with secure and trusted tools that increase user confidence in IoT environment, SOCIOTAL will enable their transition to smart neighborhood communities and cities.

CityPulse[8] provides innovative smart city applications by adopting an integrated approach to the Internet of Things and the Internet of People. The project will facilitate the creation and provision of reliable real-time smart city applications by bringing together the two disciplines of knowledge-based computing and reliability testing.

SMART-ACTION.[9] Due to the high level of interdisciplinary work in the research produced in the areas of smart cities and the Internet of Things, it will be necessary to understand, coordinate, support, and engage not only the technological elements, but also other areas such as biotechnology, social sciences, and nanotechnologies, just to name a few, that provide the right context in which Internet of Things concepts can be embedded and will be used to provide solutions that can benefit society at large.

SMARTIE[10] aims to create a distributed framework to share large volumes of heterogeneous information for the use in smart city applications, enabling end-to-end security and trust in information delivery for decision-making purposes following data owner's privacy requirements.

VITAL[11] [30] objective is the integration of ICOs among multiple IoT platforms and ecosystems. The project explores the convergence and federation of multiple IoT platforms by taking account of the cost efficiency of the deployments. In the context of VITAL, an important key factor is represented by the *virtualization* of

[7]http://sociotal.eu.

[8]http://www.ict-citypulse.eu.

[9]http://www.smart-action.eu.

[10]http://www.smartie-project.eu.

[11]http://vital-iot.eu.

interfaces that in combination with cross-context tools that enable the access and management of heterogeneous objects supported by different platforms and managed by different administrative stakeholders [31, 32].

As we shown in Fig. 4, the data and services' access of the heterogeneous objects involved in VITAL is based on the implementation of the VUAIs (Virtualized Universal Access Interfaces) that makes possible to consider a single virtual access by making the architecture platform-agnostic. The VUAI layer is built upon a so-called meta-architecture and migration layer and includes several connectors to communicate and interconnect different IoT platforms and clouds. In practice, this module deals with issues related to the management of the overall VITAL infrastructure built on the top of existing IoT architectures and cloud platforms and enables heterogeneous mashup. The VUAIs allow the implementation of a kind of abstraction, where "objects" handle that point to physical items, can be discovered, selected and filtered and also allocated.

VITAL also includes a data store for data like geographical information and smart city stakeholders. Of course, it is expected that the management of this kind of information giving location awareness and other context-related information can be effectively exploited in the optimization of computing and sensing of the management of the various clouds.

It is worth outlining that VITAL is based on W3C SSN ontology [33] that is considered ideal as a basis for unifying the semantics of different IoT platforms, since it is domain independent and extensible. Several additional concepts have to be considered to enhance the ontology starting from information about city-wide, stakeholders, IoT system, etc. The ontology update with additional functionality will allow the migration of smart city application across different urban environments.

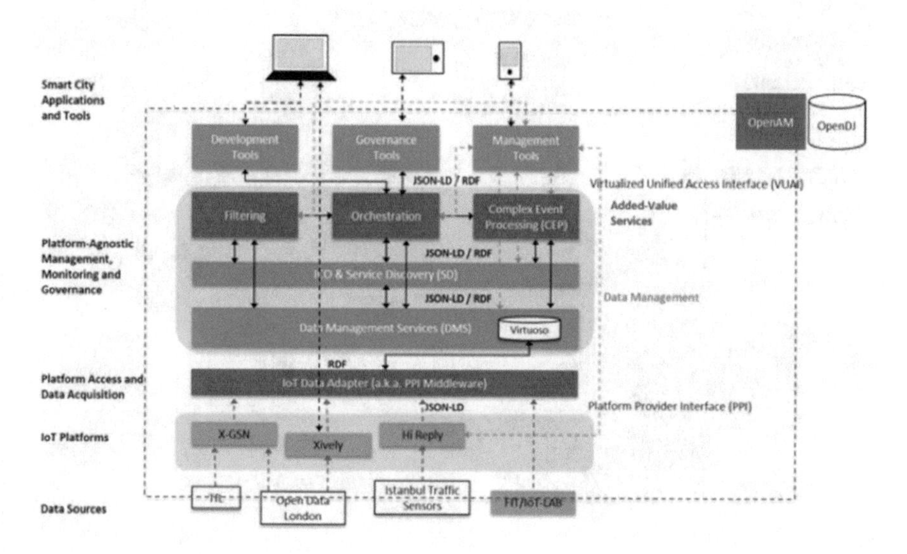

Fig. 4 VITAL platform

3 Smart Cyber-City

A *smart cyber-city* is a city where the urban environment is an integration between people, processes, places, and technologies. This ecosystem is able to self-organize and to react in order to adapt itself to the surrounding situations.

Different can be the scenarios within this context; Libelium, [34], lists 50 sensors-based applications for a smarter world (Fig. 5), to name a few, air pollution, quality of shipment conditions, smartphones detection, radiation levels, traffic congestion, water quality, waste management, smart parking, electromagnet levels, smart roads, smart lighting, noise urban maps, vehicle auto-diagnosis, and intelligent shopping.

In most urban settings, the *town center* can be considered as the core of major socioeconomic activities including tourism, social, business, shopping, work, traveling hubs (bus or train stations), education (colleges/universities). Typically, a number of people commute to the town center at different times on weekdays (mostly for work) and over weekends (shopping, work, leisure, etc.) and act as a stimulus to the socioeconomic development. These activities create an environmental track that requires to be managed in a smart way in order to migrate any negative effects on these town centers. For such purpose, the *smart city* needs a core requirement of pervasive, interconnected communication infrastructure, and access to contextual information of its citizens and physical spaces by data sensing,

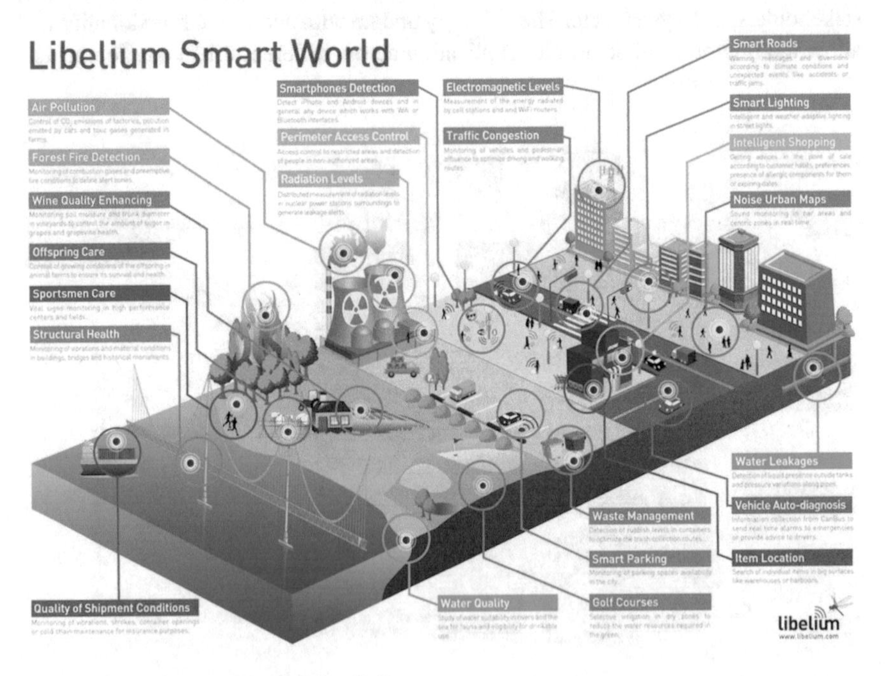

Fig. 5 Libelium Smart World vision [34]

processing and useful information for different stakeholders for consumption and decision-making [35].

For instance, a common scenario can be location of empty parking spaces during peak hours using smartphones. Another scenario relies on the information about number of vehicles available on specific roads of town center which can be detected by specific counting sensors and/or vehicle fitted with GPS devices. This information helps in route optimization by providing a traffic congestion map in real-time to the citizens who are planning to visit/leave the city and assist their decision-making toward their mode of transport (private, public) and route planning. Other scenarios concern security applications, road management, waste management etc.

Interesting technologies and solutions are introduced and discussed in the literature.

Dimitrakopoulos [36] envisions the *Internet-Connected Vehicles* concept, in which vehicles and objects of the transportation infrastructure are connected through an all IP-based infrastructure capable of exchanging information directly or indirectly and appropriate for resolving several kinds of issues, so as to result in a more efficient, safe, and green world of transportation.

In [37] Dagher et al. propose *Ubiquitous Navigation System* (UNS), a WSN-based navigation system which takes benefit from the smart street lightning system to provide a local navigation service. The idea of the authors is to make use of the already deployed WSN infrastructure for smart street lightning to provide a GPS-like service to the vehicles in the city. This WSN-aided navigation becomes important especially in cluttered and urban environments or undergrounds where GPS reception fails, thus providing a continuous smart navigation service to the user.

Crowd sourcing consists of outsourcing tasks to a "*crowd*" in an attempt to collaboratively completing tasks quickly. With smartphones becoming increasingly more powerful in terms of resources, and fitted with a variety of sensors such as GPS, gyroscopes, accelerometers, and compasses, it is possible to enable a variety of crowd sourcing applications [4]. MobSens [38], for instance, is a crowd sourcing application that can monitor air pollution and the noise levels.

The above solutions strengthen the vision toward the development of a smarter city. However, we believe that in order to realize the smart cyber city concept, those applications need to communicate and interact; CPS can play a key role as enabler of this synergy between the different actors.

3.1 Smart City Pilots

The smart city concept is still far from being realized, due to its highly autonomous and intelligent features. Indeed, the available technology is not yet sufficiently mature for smart cities to be truly autonomous, despite the recent breakthroughs in technology. However, there are currently pilot projects of smart cities in development worth mentioning.

SmartSantander [39] is a pilot project which has recently received significant attention. This is unique in the world city-scale experimental research facility in support of typical applications and services for a smart city. The test-bed deployed has dual purpose; one allows real-world experimentation on IoT-related technologies (protocols, applications, etc.), the second is supporting the provision of smart city services aimed at enhancing the quality of life in the city of Santander. Different devices have been installed: *fixed sensor nodes* attached to public lamp-posts or to building façades, which can observe a wide range of physical magnitudes (e.g., light intensity, noise, carbon monoxide, air temperature, relative humidity, solar radiation, atmospheric pressure, soil temperature, wind direction, etc.); *mobile sensor nodes* installed on the top of public transport buses, taxis, and other municipal services vehicles; *parking monitoring sensor nodes* buried under the asphalt on outdoor parking places; *traffic monitoring sensor nodes* buried under the asphalt on the main entrances to the city; *QR and NFC tags* located at city Points of Interest (POI) (e.g., monuments, bus stops, local administration premises, shops, etc.) which represent a key asset for the augmented reality service.

The development of Songdo [40] in South Korea, started from the scratch in 2001 and is predicted to be complete by 2018. Smart systems in every building are used to monitor the water and electricity, which also allow residents to connect remotely using their smartphones. Sensing technologies include RFID tags on vehicles which send signals to sensors on the road to monitor traffic flow, surveillance systems as well as smart street lights, which can be adjusted to pedestrian traffic.

Amsterdam has seen some recent developments en route to make it a smart city. The main operator of this development is Amsterdam smart city project,[12] started in 2009 with the main goal of increasing green growth using technology. Different initiatives started in specific city locations, like for example an intelligent electricity network (Smart Grid) developed in the New West district, which promises to reduce the number and duration of power outages; to improve the opportunity to feed consumer-produced electricity back to the grid; to increase capability to support the integration of electric-powered vehicles, etc.

Other important initiatives involve Barcelona.[13] Sensors have been deployed in garbage bins, allowing remote monitoring of the content of bins and optimizing garbage collection service. Additionally, the city council installed smart water system for telemanaging the irrigation of the city's green spaces, an initiative that is as good for the environment as it is for the economical aspect, with its use of latest generation technology for better resource management. Sensors on street lights detect presence and adjust the light intensity accordingly as well.

[12] http://amsterdamsmartcity.com.

[13] http://smartcity.bcn.cat/en.

London and in particular Camden Town, is one of the two cities involved in the VITAL project. Different sensors and data feeds are already available in Camden, like *cameras* used for monitoring traffic, public transport, and safe shopping; *GPS and location sensors* used for deriving information about the positioning and the status of waste collection vehicles; *meteorological data feeds* used to provide relevant information to tourists; *traffic data streams* used to balance load on public and private transport; *security data streams* used for monitoring as per national policy directives; *disaster prevention data streams* used for monitoring as per disaster plans and simulations. The aims of Camden is to boost targets of its business strategy like reducing the cost of business operations, with a view to make it an attractive destination of commercial, retail, and leisure activities; strengthen Camden's links to neighboring areas/districts (such as Euston and Kings Cross), and so on.

The other city partner of VITAL is Istanbul. As one of the largest cities in the world, Istanbul has a population of 13.9 million on a surface area of 5313 km^2. This cosmopolitan and historic city needs to meet the challenge of maintaining transportation safety and accessibility since it continues to become an important international metropolis with increasing traffic numbers. To ensure effective and efficient use of the current main arterial road network, traffic management in Istanbul is a critical issue. To deal with those issues, the idea is to develop, install, maintain, and operate intelligent transportation systems and their infrastructure including traffic monitoring and supervision cameras, radar detectors, sensors, and so on.

The Expo Milano 2015[14] is announced to be full of digital technologies that will make the overall experience easier. The visitor can use dedicated services through the multimedia Totem or smartphones that will guide them during the visit and that can be personalized according to the person's interests or available time. The smart city experience is not just limited to the exposition site but can also be used when visiting the city of Milan.

The Sunrise Lille project aims to transform the Lille1 University[15] environment into a semi-scale pilot testbed for the research, development, field assessment, and large-scale demonstration of innovative smart urban networks (SUN) for upgrading strategic and operational management capabilities of metropolitan infrasystems as water, energy, and heating supply. The major advantage of Lille1 project is to bring together governmental agencies, industry and academia, and customers.

Starting in 2012, the Sense-City project[16] will offer a suite of high-quality facilities for the design, prototyping and performance assessment of innovative, micro- and nanotechnology-based sensors devoted to urban instrumentation. Acknowledging the shortcomings of evaluating sensors performances in laboratory conditions only or in the ever-changing environment of our cities, Sense-City will provide a realistic urban test space in climatic conditions, far more complex than

---

[14]http://www.expo2015.org.

[15]http://www.univ-lille1.fr.

[16]http://www.sense-city.univ-paris-est.fr.

Table 1 Smart city projects

City/project	Purpose
Santander, Spain	Testbed platform; smart parking systems; environmental monitoring; traffic monitoring
Songdo, Korea	Smart buildings; tags on vehicles; sensors on the road; smart lighting
Amsterdam, the Netherlands	Smart grid; smart energy management
Barcelona, Spain	Smart garbage bins; smart water systems; smart parking; smart lighting
Camden Town, England	Traffic data; meteorological data; disaster prevention
Istanbul, Turkey	Smart traffic management
Milan, Italy	Augmented reality service; smart services
Lille, France	Testbed platform; upgrade strategic and operational management capabilities of metropolitan infra-systems
Sense-City Project	Testbed around mini-city concept; urban sustainability (quality of air, water, etc.)

clean rooms and far less complex than actual cities. Sense-City revolves around the mini-city concept, a large, fully customizable climatic hall able to host full- and reduced-scale models of essential urban components. The design of the models will allow for the simulation in climatic conditions of numerous scenarios of sustainable cities. The scenarios to be implemented will correspond to different research topics related to urban sustainability: energy performances in buildings, quality of air, water, and soils, quality of fluid distribution networks (gas, sewage, drink water), control of waste disposal areas, durability, and safety of infrastructures.

Table 1 summarizes the goals of the cities/projects discussed above.

4 Conclusion

This chapter focuses on the concept of cyber-physical objects and how they can help toward the realization of real smart cities capable to ensure sustainability and efficiency. Specifically, we survey the requirements and challenges in terms of *networking* and *computing* for cyber-physical system, highlighting that lightweight technologies, which enable cross-domain end-to-end communication among objects, together with cloud computing techniques can play a primary role in the field. We presented some of the most representative solutions capable to manage CPO, underlining that a lot of new enhancements are still needed in order to realize platforms ready to manage complex systems such as a smart city. In order to fill this gap, the European Commission has funded different projects under the call smart city of the Seventh Framework Programme for Research and Technological Development. We browsed the most interesting proposal, together with some of the smart city pilots already developed across the world.

The above solutions represent the first step toward the realization of the smart-cyber city concept, but still more integration is required between people, processes, places, and technologies. Only then, this ecosystem will be able to self-organize and to react in order to adapt itself to the surrounding situations.

Acknowledgments This work is partially supported by CPER Nord-Pas-de-Calais/FEDER DATA and by the European Community in the framework of the VITAL FP7 project (Virtualized programmable InTerfAces for smart, secure, and cost-effective IoT depLoyments in smart cities under contract number FP7-SMARTCITIES-608662. The authors acknowledge help and contributions from all partners of the project.

References

1. Mundi, I.: World demographics profile. Available at http://www.indexmundi.com/world/demographics_profile.html. (2012)
2. Fast Co. Design: By 2050, 70 % of the World's Population Will Be Urban. Is That A Good Thing? http://www.fastcodesign.com/1669244/by-2050-70-of-the-worlds-population-will-be-urban-is-that-a-good-thing (2012)
3. Batty, M., Axhausen, K., Fosca, G., Pozdnoukhov, A., Bazzani, A., Wachowicz, M., Ouzounis, G., Portugali, Y.: Smart Cities of the future. Eur. Phys. J. Spec. Top. **214**, 481–518 (2012)
4. Hancke GP, Bruno de Carvalho E. Silva, Hancke GP (2013) The role of advanced sensing in smart cities. Sensors (Basel, Switzerland) **13**, 393–425 (2013)
5. Aloi, G., Bedogni, L., Di F, M., Loscri, V., Molinaro, A., Natalizio, E., Pace, P., Ruggeri, G., Trotta, A., Zema, N.R.: STEM-Net: an evolutionary network architecture for smart and sustainable cities. Trans. Emerg. Telecommun. Technol. **25**(1), 21–40 (2014)
6. Lee, E.A.: Computing foundations and practice for cyber-physical systems: a preliminary report. Electrical Engineering, pp. 1–27 (2007)
7. Wu, F.-J., Kao, Y.-F., Tseng, Y.-C.: From wireless sensor networks towards cyber physical systems, pp. 397–413 (2011)
8. Banerjee, P., Friedrich, R., Bash, C., Goldsack, P., Huberman, B.A., Manley, J., Patel, C., Ranganathan, P., Veitch, A.: Everything as a service: powering the new information economy. Computer **44**(3), 36–43 (2011)
9. Basagni, S., Conti, M., Giordano, S., Stojmenovic, I.: Mobile Ad Hoc Networking. Wiley, London (2004)
10. Akyildiz, I.F., Su, W., Sankarasubramaniam, Y., Cayirci, E.: A survey on sensor networks. IEEE Commun. Mag. **40**, 102–105 (2002)
11. Pfisterer, D., Romer, K., Bimschas, D., Kleine, O., Mietz, R., Truong, C., Hasemann, H., Kröller, A., Pagel, M., Hauswirth, M., Karnstedt, M., Leggieri, M., Passant, A., Richardson, R.: SPITFIRE: toward a semantic web of things. IEEE Commun. Mag. **49**(11), 40–48 (2011)
12. Dunkels, A: Full TCP/IP for 8-bit architectures. In: Proceedings of the 1st international conference on Mobile systems, applications and services (MobiSys) (San Francisco, California, USA), pp. 85–98 (2003)
13. Kushalnagar, N., Montenegro, G., Culler, D.E., Hui, J.W.: *Transmission of IPv6 packets over IEEE 802.15.4 networks, RFC 4944 (Proposed Standard)*, Tech. report, IETF (2007)
14. Hui, J.W., Culler, D.E.: IP is dead, long live IP for wireless sensor networks. In: Proceedings of the 6th ACM conference on embedded network sensor systems (SenSys) (Raleigh, North Carolina, USA), pp. 15–28 (2008)

15. Costantino, L., Buonaccorsi, N., Cicconetti, C., Mambrini, R.: Performance analysis of an LTE gateway for the IoT. In: Proceedings of the international IEEE symposium on a world of wireless, mobile and multimedia networks (WoWMoM) (San Francisco, California, USA), pp. 1–6 (2012)
16. Zhou, M., Zhang, R., Zeng, D., Qian, W.: Services in the cloud computing era: a survey. In: Proceedings of the 4th international universal communication symposium (IUCS) (Beijing, China), pp. 40–46 (2010)
17. Patidar, S., Rane, D., Jain, P.: A survey paper on cloud computing. In: Proceedings of the 2nd international conference on advanced computing and communication technologies (ACCT) (Rohtak, India), pp. 394–398 (2011)
18. Elgazzar, K., Hassanein, H.S., Martin, P.: DaaS: cloud-based mobile Web service discovery. Pervasive Mobile Comput **13**, 67–84 (2014)
19. Costa, P., Migliavacca, M., Pietzuch, P., Wolf, A.L.: NaaS: network-as-a-service in the cloud. In: Proceedings of the 2nd USENIX conference on hot topics in management of internet, cloud, and enterprise networks and services (Hot-ICE) (California, USA) (2012)
20. Perera, C., Zaslavsky, A., Christen, P., Georgakopoulos, D.: Sensing as a service model for smart cities supported by internet of things. Eur. Trans. Telecommun. **25**(1), 81–93 (2014)
21. Zouganeli, E., Svinnset, I.E.: Connected objects and the internet of things—A paradigm shift. In: Proceedings of the international conference on photonics in switching (PS) (Pisa, Italy) (2009)
22. Kortuem, G., Kawsar, F., Fitton, D., Sundramoorthy, V.: Smart objects as building blocks for the Internet of things. IEEE Internet Comput. **14**(1), 44–51 (2010)
23. Ashton, K.: That 'Internet of Things' thing. RFiD J **22**, 97–114 (2009)
24. Vermesan, O., Friess, P., Guillemin, P., Sundmaeker, H., Eisenhauer, M., Moessner, K., Le Gall, F., Cousin, P.: Internet of things strategic research and innovation agenda. In: Internet of things—converging technologies for smart environments and integrated, pp. 7–152 (2013)
25. Hauswirth, M., Dennis, P., Decker, S.: Making internet-connected objects readily useful. In: Proceedings of the interconnecting smart objects with the internet workshop (Prague, Czech Republic) (2011)
26. Bizer, C., Heath, T., Berners-Lee, T.: Linked data-the story so far. Int. J. Semant. Web Inf. Syst. **5**, 1–22 (2009)
27. Hayes, P.: RDF Semantics. http://www.w3.org/TR/2004/REC-rdf-mt-20040210/. pp. 1–45 (2004)
28. Fortino, G., Guerrieri, A., Russo, W.: Agent-oriented smart objects development. In: Proceedings of the 16th international IEEE conference on computer supported cooperative work in design (CSCWD) (Wuhan, China), pp. 907–912 (2012)
29. Yuriyama, M., Kushida, T.: Sensor-cloud infrastructure physical sensor management with virtualized sensors on cloud computing. In: Proceedings of the 13th international conference on network-based information systems (NBiS) (Gwangju, Korea), pp. 1–8 (2010)
30. Petrolo, R., Loscri, V., Mitton, N.: Towards a smart city based on cloud of things. In: Proceedings of the international ACM MobiHoc workshop on wireless and mobile technologies for smart cities (WiMobCity) (Philadelphia, USA) (2014)
31. Petrolo, R., Loscri, V., Mitton, N.: Towards a Cloud of Things Smart City. IEEE COMSOC MMTC E-Letter **9** (2014)
32. Petrolo, R., Loscri, V., Mitton, N.: Towards a smart city based on cloud of things, a survey on the smart city vision and paradigms. Trans. Emerg. Telecommun. Technol. (ETT) (2015)
33. Compton, M., Barnaghi, P., Bermudez, L., Garca-Castro, R., Corcho, O., Cox, S., Graybeal, J., Hauswirth, M., Henson, C., Herzog, A., Huang, V., Janowicz, W.K., Kelsey, D., Le Phuoc, D., Lefort, L., Leggieri, M., Neuhaus, H., Nikolov, A., Page, K., Passant, A., Sheth, A., Taylor, K.: The SSN ontology of the W3C semantic sensor network incubator group. Web Seman. Sci. Serv. Agents World Wide Web **17**, 25–32 (2012)
34. Libelium: 50 sensor applications for a smarter world. http://www.libelium.com/top_50_iot_sensor_applications_ranking

35. Khan, Z., Anjum, A., Liaquat Kiani, S., Cloud based big data analytics for smart future cities. In: Proceedings of the 6th international IEEE/ACM conference on utility and cloud computing (UCC) (Dresden, Germany), pp. 381–386 (2013)
36. Dimitrakopoulos, G.: Intelligent transportation systems based on internet-connected vehicles: Fundamental research areas and challenges. In: Proceedings of the 11th international conference on ITS telecommunications (ITST) (Saint Petersburg, Russia), pp. 145–151 (2011)
37. Dagher, R., Mitton, N., Amadou, I: Towards WSN-aided navigation for vehicles in smart cities: an application case study. In: Proceedings of the international IEEE PerCom workshop on pervasive systems for smart cities (PerCity) (Budapest, Hungary), pp. 129–134 (2014)
38. Kanjo, E., Bacon, J., Roberts, D., Landshoff, P.: MobSens: making smart phones smarter. IEEE Pervasive Comput. **8**(4), 50–57 (2009)
39. Sanchez, L., Muñoz, L., Antonio Galache, J., Sotres, P., Santana, J.R., Gutierrez, V., Ramdhany, R., Gluhak, A., Krco, S., Theodoridis, E., Pfisterer, D.: SmartSantander: IoT experimentation over a smart city testbed. Comput. Networks **61**, 217–238 (2014)
40. Strickland, Eliza: Cisco bets on South Korean Smart City: Songdo aims to be the most wired city on Earth. IEEE Spectr. **48**, 11–12 (2011)
41. WWF, Living Planet Report, Tech. report, 2012

Structuring Communications for Mobile Cyber-Physical Systems

Luis Almeida, Frederico Santos and Luis Oliveira

Abstract Mobile autonomous agents, particularly robots, are becoming commonplace in many application domains, e.g., search and rescue, demining, agriculture and surveillance. These so-called Mobile Cyber-Physical Systems (M-CPS) allow relocating sensors and actuators dynamically in the environment to improve some global performance metric. However, the necessary agents cooperation is hindered by their heterogeneity, dynamic communication links and network topology, and by an error-prone communication channel. This chapter focuses on the networking and middleware support for a class of M-CPS and proposes the Reconfigurable and Adaptive TDMA protocol to handle data transmission under communication-related uncertainties, and the RTDB shared memory middleware to provide seamless data access across the cooperating agents. This combination of network protocol and middleware layer is particularly suited to support state sharing among agents in dynamic teams. Several experimental results confirm the advantage over other potential options. Moreover, this protocol and middleware combination have been thoroughly validated in demanding operational scenarios namely in robotic soccer teams from RoboCup Middle-Size League, which exhibit the typical requirements and constraints of M-CPS.

L. Almeida (✉) · L. Oliveira
Instituto de Telecomunicações, Universidade do Porto, Porto, Portugal
e-mail: lda@fe.up.pt

L. Oliveira
e-mail: loliveira@fe.up.pt

F. Santos
DEE/ISEC—Instituto Politécnico de Coimbra, Coimbra, Portugal
e-mail: fred@isec.pt

A. Guerrieri et al. (eds.), *Management of Cyber Physical Objects in the Future Internet of Things*, Internet of Things,
DOI 10.1007/978-3-319-26869-9_3

1 Introduction

There are many application domains where mobile autonomous agents, particularly robots, can play a desirable role, e.g., freeing human beings from repetitive or dangerous tasks or carrying out tasks in ways that are much more efficient than humans could possibly do. Examples of such domains range from search and rescue, demining, agriculture, surveillance and industrial production. In many cases, it is desirable to use multiple cooperating agents, for example to improve performance by having several agents working in parallel to speed up completing the main task, as in area coverage. In other cases, it is even mandatory to use multiple cooperating agents such as when the task requires physical capabilities that go beyond what a single agent can provide, e.g., transporting large items.

Teams of mobile autonomous agents are a special case of Cyber-Physical Systems in which the intertwining between the cyber and physical worlds is flexible. In fact, multiple agents imply a diversity of sensors and actuators that can be relocated dynamically. We call these systems Mobile Cyber-Physical Systems (M-CPS).

M-CPS also imply that the agents in the team cooperate towards a common goal. Cooperation in Robotics, is an old research topic [1] and it can be carried out at multiple levels, e.g. cooperative perception, cooperative learning or cooperative planning and execution [2]. Overall, the purpose of cooperation is to improve global performance through either improved sensing, improved control or more efficient actuation or use of resources.

However, achieving cooperation is not trivial given that agents are frequently heterogeneous, they set up and break communication links leading to a highly dynamic network topology, and the communication channel is prone to interference from many sources. Therefore, an appropriate framework is needed to support the development, deployment and execution of cooperative behaviors while making such cooperation efficient and effective.

In this chapter we address two fundamental dimensions of such framework, namely communications for data transmission and middleware for data access. We provide solutions for both that have been successfully used in several applications, particularly in robotic soccer teams from RoboCup Middle Size League (MSL) [3]. These are also M-CPS exhibiting their typical requirements and constraints.

Concerning the communications, we briefly present the Reconfigurable and Adaptive TDMA protocol [4] that has been implemented on top of widely used Commercial Off-The-Shelf (COTS) wireless communication technologies. It allows obtaining gains in terms of packet losses and delays while adding synchronization and membership services, coping with agents that dynamically join and leave the team as well as with alien traffic potentially generated by neighboring nodes not engaged in the protocol.

On the middleware level we present the Real-Time Database (RTDB) [5] shared memory middleware that provides a transparent access to remote data with age information. It is particularly suited for sharing state among the team members, which is the basis for many cooperative behaviors.

The remainder of this chapter is organized as follows. The following section discusses architectural aspects of M-CPS and defines the minimum communication-related requirements of our target architecture. Section 3 briefly surveys communication technologies and middleware layers and establishes the need for our proposals. Section 4 presents the RTDB middleware while Sect. 5 describes the Reconfigurable and Adaptive TDMA protocol, including a performance assessment. Section 6 concludes this chapter.

2 Architectural Aspects of M-CPS

In this section we briefly analyze the architecture of M-CPS concerning their physical topology on one hand, and basic services for cooperation on the other. In [6] we find a thorough discussion of M-CPS and their architecture.

2.1 *Physical Topology*

M-CPS, as defined previously, are composed of several autonomous agents that cooperate towards a common goal. Physically, this implies a certain number of nodes, i.e., the agents, and their inter-connections, thus a network.

From small teams of a few agents [7] to swarms of many agents [8], M-CPS come in many different shapes and sizes. Despite that, due to practical reasons related to cost, deployment, and operation, the former are more common.

The mobility requirements of M-CPS imply that communications must use a wireless medium, tolerate non-line-of-sight conditions as well as frequent link disruption and set up. Nonetheless, the specific kind of wireless communication technology must be adapted to the environment in which the team operates, leading to different restrictions in terms of reliability, bandwidth, and communication range. For example, rf communication is typically used over the air while (ultra-)sound is used within aquatic environments [9].

By restricting the team size, certain technical options become viable, such as using global information and centralized topology. A commonplace solution within small teams is the star topology in which all the communications go through a central repeater that can be fixed in the environment or installed on one of the agents. This topology limits the diameter of the network but simplifies the dissemination of information, thus achieving consensus, since all agents can only receive information from the repeater [4]. Conversely, the mesh topology has the advantage of allowing to extend the reach of the team without losing connectivity [10] but complicates achieving consensus. Meshes also extend naturally to swarms.

However, swarms can also be clustered in multiple stars in which the repeaters become cluster heads that also handle the inter-cluster communications.

Nevertheless, one important aspect to retain is that, independently of the size, the number of nodes will be variable as agents start and stop asynchronously, and move in and out of communication range, translating to a dynamic topology.

2.2 Basic Services

The proper operation of M-CPS relies on top of several basic services that must be provided by a communication protocol and middleware. These are part of the system functional architecture:

- **Synchronization**. Efficiently coordinating a team of autonomous agents requires some sort of synchronization. This can be relevant for communications, formation control, etc. Two kinds of synchronization are possible, clock synchronization that provides a common global notion of time [11], or logical synchronization that simply provides a consistent sequencing of events [12]. Despite the former being more powerful it requires a specific clock synchronization service, while the latter can generally be inferred just from the data communications.
- **Information dissemination**. Cooperation typically implies some sort of communication of either events or state. This is information that must be disseminated through the network and it is time constrained in both cases. In a star topology, this can be achieved efficiently with broadcasting or multicasting. However, in a mesh or clustered topology the information must be diffused, typically using a flooding technique, implying longer latency.
- **Dynamic membership**. Another common requirement for cooperation is the notion of team, i.e., knowing at each instant which agents are engaged in the team. This allows dynamically readjusting the coordination approaches to better exploiting the resources currently available in the team.
- **Location-awareness**. Knowing agents location, at least with respect to each other, is also very important for most cooperative behaviors [13]. This can be achieved with information from the communication network, such as attenuation models and time-of-flight, or with other means, such as GPS and vision.
- **Combination of behaviors**. Finally, despite engaged in a team, agents still have to take decisions autonomously, particularly those related to individual safety. Autonomous behaviors are also needed whenever agents lose connection with the team. Therefore, all agents need to have a combination of coordinated with autonomous behaviors for proper operation in a wide range of situations.

2.3 Our Target Architecture

Our focus is on small teams of agents for indoor applications, such as surveillance, cleaning or entertainment. We consider that the agents communicate with a standard COTS RF technology, for the sake of availability, and in a star topology, for the sake of consistency and timeliness.

Moreover, we consider logical synchronization, multicast data dissemination and automatic membership support, to favor simplicity. We define just this set of minimum requirements to grant high versatility to our solutions. We believe, however, that these requirements are shared by a relatively broad range of applications. On the other hand, we explicitly leave localization and behaviors out of our architecture for their strong application dependency.

3 Communications and Middleware

Defining communications and middleware solutions that satisfy the minimum requirements referred before (Sect. 2.3) is, in some sense, our problem statement. To address this problem we briefly review existing options concerning communication protocols and middleware layers before we put forward our own solutions.

3.1 Communication Protocols

When restricting our choice to standard RF communication solutions that are widely available in the market, i.e., COTS, and also widely used within the Robotics community, two solutions stand out. Small teams typically involve agents that are more computing capable and typically have IEEE 802.11 RF interfaces. In the case of swarms, the agents are typically simpler and use technologies that are less energy consuming and can be embedded in smaller nodes. Probably the most used RF technology in this case, today, is IEEE 802.15.4. Thus, our choice naturally falls on IEEE 802.11, particularly in infrastructure mode, which is today the de facto standard for general purpose LANs.

The IEEE 802.11 standard in infrastructure mode already presents a star topology where the repeater is the Access Point (AP). It also provides support for multicast and broadcast transmissions for information dissemination but it still lacks synchronization and membership services.

The protocol itself has several features, added in the last few years, for improved support to Quality of Service (QoS) differentiation, the 802.11e amendments. But none of these provides a simple answer to the unmet requirements. The same happens with many overlay protocols that were developed to improve the protocol performance, from enhancements in the ap signaling to improve throughput by dynamically adapting the carrier sense threshold in the nodes [14], to modifying the back-off mechanism for reducing collisions in overload [15], to scheduling the periodic traffic from a central node [16] or even organizing the real-time transmissions in a round, in a TDMA fashion, with higher priority than the remaining traffic [17].

However, as explained in [18], to the best of the authors knowledge there was no protocol that simultaneously considered dynamic and flat team composition, while supporting the presence of uncontrolled external traffic of potentially similar priority, and without introducing hardware or device driver modifications. This ultimately led to the development of the Reconfigurable and Adaptive TDMA protocol that we present in Sect. 5.

3.2 Middleware Layers

The middleware layer complements the communication protocol providing data abstractions and data access methods that facilitate the development of cooperative behaviors across possibly heterogeneous agents. There are many such layers, normally developed to support generic distributed applications such as CORBA and DDS standardized by OMG, ICE provided by ZeroC, and SOAP standardized by W3C. All these have variants specialized for Robotics applications, such as MIRO, ORTE, ORCA or ROS, respectively. Among these, ROS has received substantial attention from the Robotics community recently, and has become widely used. Other variants exist aiming at different application scopes, e.g., Smart Objects [19], which cater for the respective requirements.

Concerning Robotics middleware, their common requirements are defined in [20]. From these we highlight the following three for their higher relevance:

- **Simplify application development** through provision of high level abstractions and simplified interfaces. This allows hiding heterogeneity and low-level communication details.
- **Support communications and interoperability**, allowing the integration of modules from different sources and their automatic discovery and configuration.
- **Provide efficient resource utilization**, e.g., processors, networks and memory.

At the end, the requirements of a middleware layer depend on the specific application domain. As stated in [21], "*(...) although the middleware solution is very useful, it is difficult to have one middleware platform that can offer all the required features and functionality for collaborative robotic systems.*" In our case, we set the following specific goals, beyond satisfying the requirements expressed above:

- **Sharing state** in periods of **high team interaction**. This goal translates to two requirements, one on low overhead in both computations and communications,

and another one on improved data timeliness, particularly in terms of quick access and age information;
- Separate **data access** from **data transmission**. This goal implies the use of data proxies, which are read/written autonomously by the network while providing asynchronous data access to the local processes.

These goals point to a thin middleware layer that provides synchronization to keep track of the end-to-end transactions latency for age computations, and that relies on the shared memory model for efficient implementation of data proxies. These proxies also have significant impact on timeliness because local processes access remote data through them. Since they are also local, there is no communication delay involved when a remote variable is accessed through its proxy. This brings along a desirable separation between computing delays and communication delays. The proxies, then, have to be properly updated in the background by an adequate communication protocol.

The result was the RTDB shared memory middleware that we present in Sect. 4. This middleware provides data access but requires a communications protocol to provide data dissemination. Despite the possibility of using different communication protocols, we use, for its good match, the aforementioned Reconfigurable and Adaptive TDMA protocol.

4 Real-Time Database

The RTDB middleware builds on the architecture of the classical Blackboard [22, 23], which was developed in the early days of Artificial Intelligence and still widely adopted in many applications. The Blackboard is a public repository of information that can be local or remote, where processes can read and write data. It can include raw input data, partial and final results, and control information. It also acts as a communication medium and buffer. Similarities and complementarities between Blackboard-based systems and multi-agent systems are discussed in [22].

The RTDB extends the Blackboard providing data interfaces (proxies) locally in all participating agents similarly to a multi-port memory. These interfaces are implemented with local data structures, one in each node, which constitute the local instances of the RTDB (Fig. 1). These are divided in two areas, depending on the scope of the data they hold:

- **Shared**—That keeps the data that is relevant for cooperative behaviors and which will be disseminated within the team. This area is organized in blocks, one dedicated to each agent. In particular, one block contains the data that the holding agent shares, which will be transmitted to the team, while the remaining blocks contain the data shared (transmitted) by the other agents;
- **Local**—Used to hold the data that is only relevant to local processes and which will not be transmitted to the other agents.

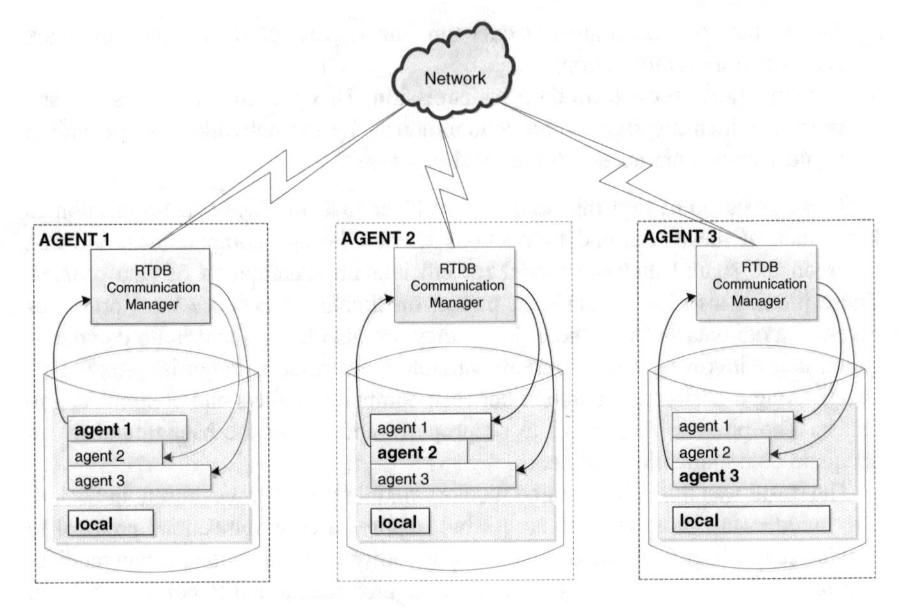

Fig. 1 Network view of the RTDB architecture

Each process connects to the RTDB through the local RTDB Application Programming Interface (API) that provides the necessary methods to access the data, transparently to the block in which the data actually resides inside the RTDB.

The heterogeneity of agents generates different requirements for memory usage leading to different RTDB block sizes. The block size and the choice of data that must be shared or kept locally is defined a priori through a configuration file.

The RTDB can be configured with both shared and local areas or with just one. Independent agents can use the local area, only, for inter-process communication but cooperating agents always need a shared area for team-wide communication. Whenever this latter area is implemented, a *communication manager* refreshes the RTDB contents in the background, ensuring consistency between the original data items and their proxies. In particular, it must consider the specific temporal validity of the data items, the constraints of the communication medium and the amount of data to exchange.

According to the specific objectives of the middleware, the communication manager can use a different protocol. In our case, considering the objectives defined in Sect. 3, particularly state sharing across the team, we consider the Reconfigurable and Adaptive TDMA communication protocol described in Sect. 5. Other refreshing policies and protocols are possible, for example to favor events dissemination instead of state sharing.

Finally, the access to the RTDB items configures a typical single writer multiple readers synchronization case. We solve data consistency issues in this situation using a well known double buffering technique due to its simplicity and reduced blocking, despite an extra cost in memory usage.

4.1 Configuration

The configuration of the RTDB is done automatically by parsing a configuration text file that specifies the team characteristics and the RTDB composition. As mentioned previously, the team can be composed of multiple agents with different roles and equipped with different sensors and actuators, thus having distinct data requirements for either local and global communication. Hence, the configuration file allows describing each agent from the data point of view and altogether represents the team model. This description is done using the following constructs:

- `AGENTS`—Set of agent unique identifiers that specify all agents that may integrate the team. Each agent identifier `(id_ag)` will also be used in the actual application code when accessing remote data in the RTDB to specify which agent to retrieve the data from;
- `ITEM`—This is a data unit kept in the RTDB and handled as an integer piece of information. Each such item is identified by a unique identifier `(id_it)` together with the following three attributes:
  - `Datatype`—This is the actual type of the item data and it is used to compute the data `size` in bytes, necessary to hold the item in memory. It can be a predefined type, such as `int` or `double`, or a user defined type;
  - `Headerfile`—This attribute is needed if the data type is user defined, only. It contains the path and name of the `C language` header file where the data type is defined;
  - `Period`—Item refresh period in multiples of the communication cycle. This allows adjusting the RTDB communication requirements to the actual dynamics of its items. It is used in shared items, only.
- `SCHEMA`—Set of local and shared `ITEMs` produced by a given agent type. Each schema has a unique identifier `(id_sc)`. There can be one or more schemas. The `ITEMs` are thus specified in two lists, accordingly:
  - `Shared`—The list of `ITEMs` to be shared, i.e., broadcast;
  - `Local`—The list of `ITEMs` that are available to local processes, only;
- `ASSIGNMENT`—Associates one `SCHEMA` to one or more AGENTS and thus allows defining all RTDB instances. There can be one or more assignments depending on the number of different agent types.

To better explain how the configuration file is used, Listing 1 shows an example of a configuration for a team of three `AGENTS` with two similar mobile robots (`robot1` and `robot2`), that explore the environment using a camera, and a base station (`base`) responsible for data fusion and world model construction.

The image acquisition processes running on the robots save the raw images data in the local item `image`. The self-localization and obstacle detection processes read the `image` item and save the results of self-position and obstacles localization in

```
AGENTS = robot1 , robot2 , base ;

ITEM image {datatype = struct image ; headerfile = image.h ; }
ITEM position { datatype = struct pos ; headerfile = pos.h ; period = 1 ; }
ITEM obstacles { datatype = struct obstacles ; headerfile = obstacles.h ;
    period = 1 ; }
ITEM fuse_data {datatype = struct fuse ; headerfile = fuse.h ; period = 1; }

SCHEMA robot { shared = position, obstacles ;
               local = image ; }
SCHEMA base_st { shared = fuse_data ; }

ASSIGNMENT { schema = robot ; agents = robot1 , robot2 ; }
ASSIGNMENT { schema = base_st ; agents = base ; }
```

Listing 1 Example of an RTDB configuration file

the shared items `position` and `obstacles`, respectively. These two items will be broadcast to the other agents every communication cycle.

A third node (`base`) equipped with a powerful computing system, not necessarily a mobile robot, is responsible for constructing the world model of the environment, combining the data received from the robots. This world model is then shared with the robots, for example, allowing them to choose areas that are still to be explored, improving the efficiency of a cooperative slam approach.

This example shows the capacity of the RTDB middleware to support cooperative applications among heterogeneous agents. However, the team composition and data requirements must be known a priori, when the code of the agents is compiled, and thus cannot be changed during execution. Note, nevertheless, that a team can be configured according to its maximum dimension and requirements. Then, at run time, the actual number of working robots can be less but the cooperative applications need to be prepared for this possibility.

4.2 RTDB API

The RTDB is fully implemented in `C language`. The functionality of the RTDB is available through a very simple API with only four methods, as shown in Listing 2. Two additional methods are used internally for updating the remote items.

```
public:
  int DB_init (void)
  void DB_free (void)
  int DB_put (int id_it, void *data)
  int DB_get (int id_ag, int id_it, void *data)

protected:
  int DB_comm_init(RTDBconf_var *rec)
  int DB_comm_put(int id_ag, int id_it, int size, void *data, int age);
```

Listing 2 The RTDB interface methods

The DB_init method is called once by every process that needs access to the RTDB and handles initialization issues. The actual memory allocation for holding the RTDB in each agent is executed by the first process to invoke such call. Subsequent calls just increment an internal process counter. Conversely, the method DB_free does the corresponding clean up and detaches a process from the RTDB. It decrements the process counter and, when zero, frees the respective memory blocks. Both DB_init and DB_free return the value 0 upon successful execution or −1 in case an error occurs.

The actual access to the RTDB memory areas is carried out with the non-blocking methods DB_put and DB_get. Both methods use the item identifier id_it defined in the configuration file as well as a pointer to the data to be written to or read from the RTDB, respectively. DB_get further requires the specification of the agent id_ag from which the item to be read belongs to, which is used to identify the respective data area in the database. This method returns the age of the data retrieved from the RTDB in milliseconds or −1 in case of error. The method DB_put returns the total number of bytes copied to the RTDB or −1 indicating an error.

The access to items defined as local or shared is transparent since the item identifier id_it is unique for all the items saved in the RTDB. This allows transforming a local item to shared, or vice versa, with a simple change in the configuration file, simply moving that item across the respective lists. This can be very useful for debugging purposes, to have temporary access to agents' local data at run time, in a monitoring station.

Finally, the RTDB API includes two protected methods (DB_comm_init and DB_comm_put) that are not available to an ordinary user process being used exclusively by the communications manager in each agent.

With respect to the RTDB access, the communications manager is rather similar to an ordinary user process, also making use of the same DB_init, DB_free and DB_get methods to begin and finish the access to the RTDB and to read data from the RTDB shared areas. However, it uses DB_comm_init, invoked once when the process is launched, to retrieve communications relevant information from the RTDB, such as period and size for all items that are to be transmitted by this agent, allowing to compute an internal transmissions schedule that determines which items to transmit in each communications cycle.

On the other hand, the DB_comm_put method is used to write in the shared RTDB areas the remote data received through the communications interface. The size parameter is used for a simple validation of the received data, comparing the received data size with the expected data size. The age parameter is the age of the received data at the reception instant, thus including the producer and transmission age components. This is an age offset that will allow computing the total data age at the time of data consumption, by the DB_get method.

4.3 Age of Data

Knowing the age of the data can be very useful for cooperative behaviors to detect and possibly mitigate situations of loss of temporal validity. However, for the sake of simplicity, our middleware does not include a global clock service implying that the clock in each robot is not correlated. To circumvent such difficulty, the middleware computes time intervals, only.

When a producer writes an item in the RTDB, the local time t_1 is saved in a field of the item record called `timestamp`, as shown in Fig. 2. Later on, when the communications manager fetches the data to disseminate it to the other agents, it computes for each item to be transmitted the difference between the current local time t_2 and the saved t_1, which is the age of each datum at the time of transmission in the producer side. The datum age at the producer side $t^p = t_2 - t_1$ is attached to each datum itself and transmitted together in the network packet.

When the packet is received by the communications manager at the consumer side, each item is individually written in the RTDB shared area that corresponds to its producer. The data age received from the producer is subtracted from the current local time at the consumer, t_3, and the result, $t_3 - t^p$, is saved in the `timestamp` field. When a consumer process retrieves the item from the RTDB, the difference from the current time to the value saved in `timestamp` is computed, resulting is an estimate of the age of the data, from the moment it was produced (inserted in the RTDB) to the moment in which it was consumed (read from the RTDB).

This estimation, however, still lacks the transmission time (T_{wt}), which depends on the actual bit rate, on the latency to access the medium and on possible re-transmissions. However, as shown in Sect. 5, the use of the Reconfigurable and Adaptive TDMA communication protocol has a positive impact on the transmission time, leading to a relatively constant latency that can be easily added to the age estimation to improve its accuracy.

4.3.1 Upper Bounding the Age of Data

For the sake of flexibility, the communication protocol is not necessarily synchronized with the control system of the robots. This may lead to extra delays in the

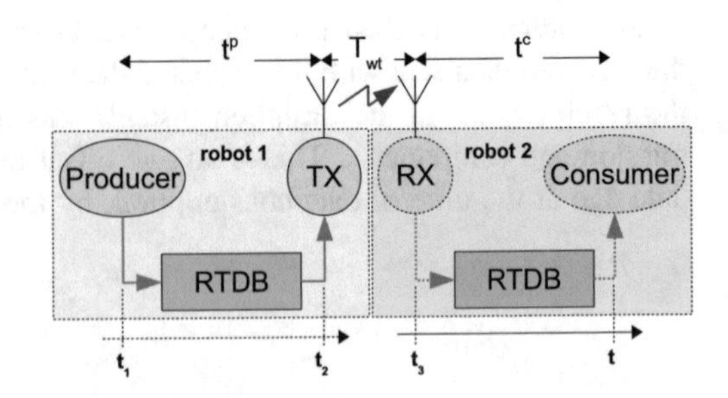

Fig. 2 Datum age calculation

refreshing of the remote data that the programmer must be aware of. In particular, when a robot accesses a local image of a datum from another team member, that datum can be as old as *max_data_age* in Eq. 1.

$$max\_data\_age = min(T_{rcpp}, n_{dup} \times T_{tup}) + T_{wt} + (n_{dup} \times T_{tup}) \quad (1)$$

This worst case data age corresponds to when the communications manager fetches the data in the RTDB for transmission just before that data being updated by the respective producer process in the respective agent. Thus, at that point, that data can be as old as one period of the respective producer (T_{rcpp}).

However, this latency cannot be larger than the item update period configured in the RTDB ($n_{dup} \times T_{tup}$) thus, the minimum of the two must be considered. Note that n_{dup} is the refresh period in integer number of communication cycles defined in the item control record, and T_{tup} is the communication cycle duration, defined in the Reconfigurable and Adaptive TDMA protocol as the Team Update Period. Moreover, the actual item refreshing is considered to be strictly periodic. The transmission of the data over the air takes some time that must also be accounted for (T_{wt}). Finally, when the consumer accesses the data on its side, the data can be waiting in the respective item buffer for at most another item update period ($n_{dup} \times T_{tup}$). Within Eq. 1, only the wireless transmission delay is unknown and may vary with the traffic load in the network, requiring an adequate estimation.

The minimum age of any datum corresponds to the situation in which the transmission takes place right after the producer updated the item and the consumer accesses the item right after it has been received, being thus given by Eq. 2.

$$min\_data\_age = T_{wt} \quad (2)$$

This large difference between maximum and minimum age shows that the item age can be affected by high jitter as typical in situations in which items are propagated through unsynchronized cycles. The order of magnitude of *max_data_age* determines the dynamics of the cooperative behaviors that this database management system can cope with.

5 Reconfigurable and Adaptive TDMA Communication

The Reconfigurable and Adaptive TDMA protocol is an overlay protocol that can be implemented over several communication technologies. Here we will consider IEEE 802.11 in infra-structured mode. The protocol structures the agents transmissions in a round allocating one slot to each agent. The agents attempt their transmissions at the beginning of their slots. The following properties are specific to this protocol:

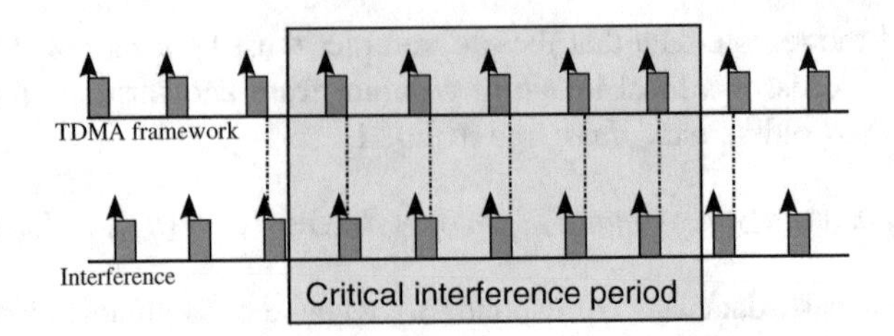

Fig. 3 Interference of a coherent periodic source on a TDMA framework

- **Constant round duration**. Apart from the round phase variations referred next, the round duration is constant, matching the real-time requirements of the conveyed data. In our case, it matches the update requirements of the data inside the RTDB and thus we refer to it as Team Update Period (T_{tup}).
- **Adaptive round phase**. Delays (δ) suffered by the agents transmissions inside their slots and within a certain bound (Δ) cause an adaptation (shift) of the round phase in an attempt to enforce the inter-transmissions separation. This mechanism allows maintaining the agents synchronized in the round without resorting to clock synchronization as usual in common TDMA protocols.
- **Reconfigurable round structure**. The maximization of the separation between transmissions is enforced, even when agents join and leave the team at run-time, by maximizing the slots duration. This is achieved dividing the round by the number of currently active team members K.

Maximizing the separation between the agents transmissions has two consequences, it reduces collisions among team members and it creates a periodic traffic pattern that is permeable to external traffic, i.e., uncontrolled traffic sent by other sources outside the team.

Moreover, the phase adaptation scheme also allows escaping from coherent periodic interfering sources. These sources, which have periods that are close to integer multiples or submultiples of the protocol period, can have a significant negative impact in network performance if the round phase is not shifted, even with low load (Fig. 3), due to persistent interference.

This pernicious phenomenon also happens when the robots in the team transmit periodically, and with similar periods, but unsynchronized. In this case, each one will be a coherent periodic interference for the others.

5.1 Round Phase Adaptation

In order to carry out the round phase adaptation, one agent is elected to be the round reference. It is such agent, only, that carries out the adaptation. All the remaining $K - 1$ agents synchronize to that reference and determine their slots adding the corresponding offsets (Fig. 4).

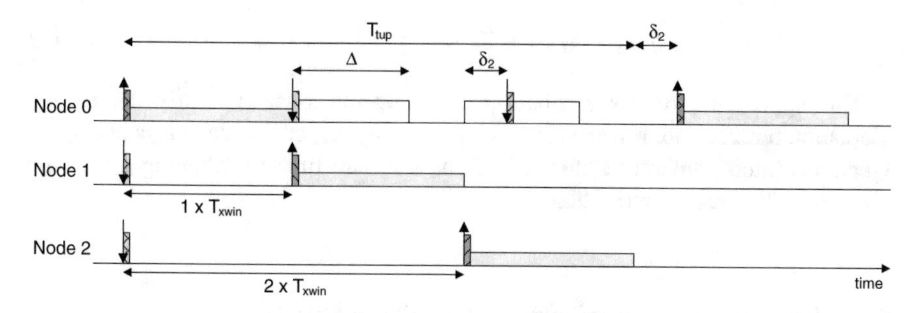

Fig. 4 TDMA round adaptation by agent 0 as reference

In all agents, the transmissions are triggered with the help of a timer. When the timer in agent *i* fires at time $t_{i,now}$, it issues its transmission and sets the timer to fire at $t_{i,next} = t_{i,now} + T_{tup}$, i.e. one round after. However, during this interval, the agents correct their next transmission instant for synchronization purposes. If, for some reason, e.g., a packet loss, such correction is not done, then they transmit strictly periodically with period T_{tup}. The corrections of the next transmission instant are different depending on whether it is the round reference agent or other agents.

The round reference agent transmits in slot 0. During the round, it continues monitoring the arrival of the packets from the other agents in the team. When the packet from agent *j* arrives, the reference agent computes the delay δ_j between the effective and expected reception instants. Its next transmission, at $t_{0,next}$, is then adjusted according to the longest such delay among all the packets received from the team in one round, as long as they are within a validity window $[0, \Delta_K]$. This is expressed in Eq. 3.

$$t_{0,next} = t_{0,now} + T_{tup} + max(\delta_j)_{j=1..K-1, 0 \leq \delta_j \leq \Delta_K} \tag{3}$$

Note that Δ_K bounds the maximum possible round phase adaptation. The effective round period will be within $[T_{tup}, T_{tup} + \Delta_K]$. The index *K* indicates that it varies with the number of active agents in the team. In fact, Δ_K is a fraction ε, configured statically, of the slot size T_{xwin_K} which in turn results from an even division of the round in *K* slots.

$$\Delta_K = T_{xwin_K} \times \varepsilon, \quad 0 < \varepsilon < 1, \quad T_{xwin_K} = \frac{T_{tup}}{K} \tag{4}$$

Larger values of ε can make the round too irregular because of too frequent adaptations while shorter values smooth the round duration due to fewer adaptations.

In what concerns the remaining agents, they will eventually receive a reference packet during the round. When they do, they estimate the instant at which the reference packet was transmitted, t_0, and they adjust their next transmission instants $t_{i,next}$ using appropriate offsets to the respective slots (Eq. 5).

$$t_{i,next} = t_0 + i \times T_{xwin_K} \quad i = 1 \ldots K - 1 \tag{5}$$

This approach provides a robust synchronization since it clearly defines the adaptation bounds and, if packet losses affect the round, or if wide delays occur, the agents continue implicitly synchronized for a certain time by adopting a periodic behavior with their current phase.

5.2 Reconfiguration of the Round Structure

The dynamic reconfiguration of the TDMA round structure allows coping with variable number of team members while maximizing the slot size. This specific mechanism supports the dynamic insertion/removal of agents in the round in a fully distributed way.

One important aspect concerns the agents identification and their assignment to slots. Each node has a unique physical ID given at configuration time. Since we cannot know which agents will be active in the team at each moment in time, we cannot assign agents to slots statically. Thus, we use a simple rule to map the static and unique physical ID to a dynamic logical ID. The currently lowest physical ID among the running agents is assigned dynamic ID 0 (and slot 0), the following physical ID is assigned dynamic ID 1 (and slot 1) and so on until the highest physical ID among the running agents that is assigned dynamic ID $K - 1$ (and slot $K - 1$). This assignment is carried out every time there is a change in the slots structure, i.e., every time an agent joins or leaves the team.

Another particularly important aspect is the consistency of the membership function across the team, which determines the number K of active team members. In fact, all agents need to compute similar values of K so that they generate compatible TDMA round structures. To enforce consistency in the reconfiguration mechanism we use a membership vector added to the packets transmitted by each node, containing its perception of the team status (Fig. 5). We define four states in terms of membership:

- **(N)ot running**—The agent is not powered up or is unreachable, i.e., not associated with the team wireless Access Point;
- **(I)nsert**—The agent started transmitting but has not yet been detected by all the current team mates. In this state the agent has no slot yet in the TDMA round and thus it is transmitting out of phase as external traffic;
- **(R)unning**—The agent has been detected by all team members and it is transmitting in its own slot in the TDMA round.
- **(D)elete**—The agent is not transmitting or its message was not received, e.g., due to an error.

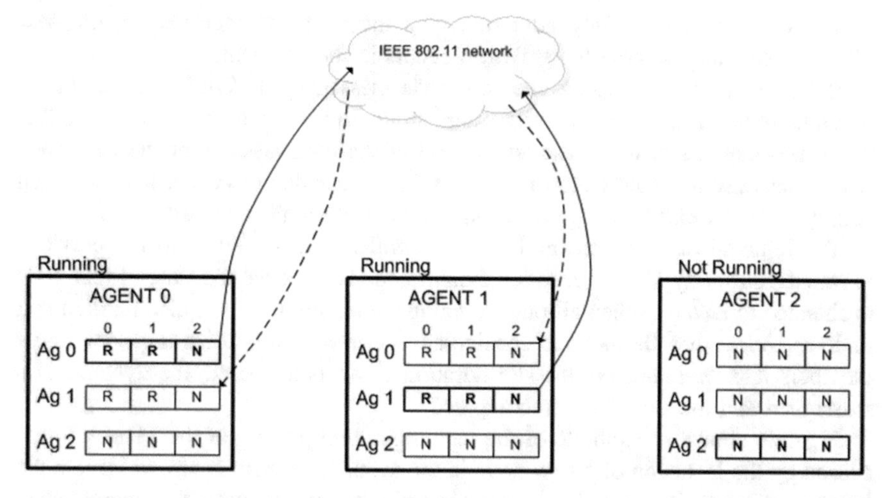

Fig. 5 Dissemination of the membership vectors

5.2.1 Reconfiguration Process

When a new node arrives, it starts transmitting its periodic information in an unsynchronized way, i.e., as external traffic, with its own state as *Insert*. Meanwhile, all nodes, including the new one, continue updating their membership vectors with the received frames. During this initial period called *agreement* phase, the new node has no slot in the round, thus no dynamic ID, and the T_{xwin_K} value remains unchanged.

At the instant of its following transmission, one round later considering no errors, the new node checks if all current team members have agreed on its presence, i.e., they all detected its transmissions and signaled it marking the new node in their membership vectors as *Insert*. This will eventually occur, leading to the end of the agreement phase and the start of the *reconfiguration* phase.

In the reconfiguration phase, the new node updates the number of active team members K, the slot duration T_{xwin_K}, reassigns the dynamic IDs locally, computes the offset of its slot in the new round configuration, updates its state to *Running* and transmits its packet, still unsynchronized. Upon the reception of the next reference packet (dynamic ID 0), the slot offset is used to set a timer and trigger the new node transmission in its new slot thus concluding the integration of the new node.

However, the complete reconfiguration process only ends when all nodes adhere to the new round configuration. This is carried out node by node, as their transmission instants occur. Basically, every time a transmission instant fires, the respective team member checks whether any new node has joined during the past round and has been acknowledged by all other members, i.e., it already transmitted with *Running* status. If it has, then this node also reconfigures the round updating K and T_{xwin_K}, reassigning the dynamic IDs and computing the offset of its slot in the new round configuration. Then it transmits its packet, which is still in the slot of the previous configuration. Only after receiving the next reference packet, the new offset is used and the next transmission occurs in the new slot.

If this node did not receive the new node message with *Running* status in the previous round, it keeps the current round structure for one round more. Note that this can occur due to the phase adjustment of the transmissions of the new node when transiting to its new slot, but it can also occur due to packet losses, which simply cause an extension of the reconfiguration phase for an extra round.

The removal of an absent node uses a similar process. When in the previous *m* rounds, currently 10, no reception from a node is detected, the state of that node is changed to *Delete*. When all other running team members have also marked that node as *Delete* then the node is considered as *Not Running*, the number of active members *K* is decremented, the slot duration T_{xwin_K} is increased, the dynamic IDs reassigned and the slot offsets recomputed.

Figure 6 shows an example of the reconfiguration process of the TDMA round caused by the inclusion of a new node in the team. Note that the arrows denote the instants of transmission and the following blocks are the respective membership vectors (representation not at scale).

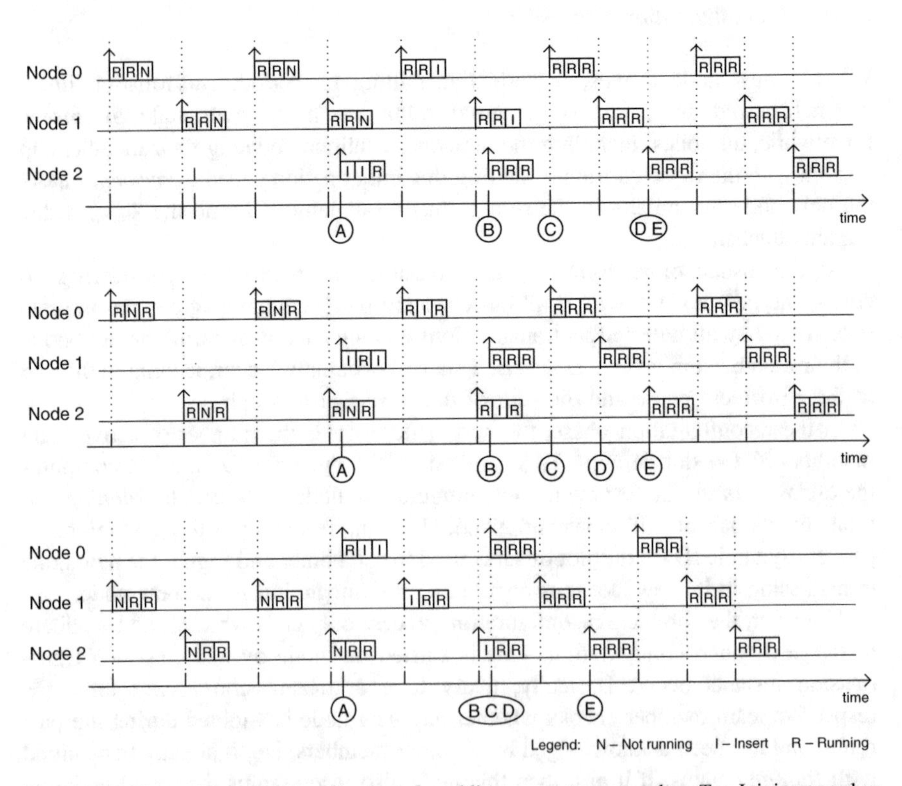

Fig. 6 Timelines of three joining situations for adding a new team member. *Top* Joining member has highest ID; *Middle* Joining node has an intermediate ID; *Bottom* Joining node has the lowest ID and will become the new reference

In the upper timeline in Fig. 6, the joining node has the highest physical ID. The agreement phase is delimited by A and B. The new node is integrated at D, which also corresponds to the end of the reconfiguration phase. In the middle timeline, the joining node has an intermediate physical ID with respect to the nodes already in the team. In this case, the node will be integrated in the round and the reconfiguration phase ends one slot after. Finally, in the lower timeline the joining node has the lowest physical ID with respect to the nodes already in the team and thus it will become the new reference. Note that once it changes its state to *Running*, it reconfigures the round internally and immediately becomes the new reference node with dynamic ID 0. Thus, it is integrated as soon as the agreement phase ends, and all the other nodes will recompute their offsets with respect to its transmission. The reconfiguration phase ends with the last slot of this round.

The relevant points in the reconfiguration process timeline, namely A through E, are explained next.

- **A**—A new node has connected to the network and, after waiting T_{tup}, it starts transmitting, unsynchronized;
- **B**—After one round, the new node has received messages from all the other nodes with their membership vectors indicating the new node as *Insert*. Thus, the agreement phase is over and the reconfiguration phase is started. In its membership vector, it updates all states to *Running*, increments the number of team members to $K = 3$, updates the slot duration to $T_{xwin_K} = T_{tup}/3$, computes the offset of its slot and transmits its packet. Upon reception of this packet, all the other nodes update the state of the new node to *Running* and perform a similar round reconfiguration to 3 slots;
- **C**—The next reference packet is transmitted and received by all nodes. Upon this reception, each node sets up a timer to trigger the respective packet transmission in the right slot;
- **D**—The timer of the new node expires and this node transmits its packet in its newly allocated slot;
- **E**—The last node to transmit its packet in the new round configuration ends the reconfiguration phase.

5.2.2 Time to Join the Team

We define the latency of a new node i joining a team with $K - 1$ members (K upon joining and $i = 0 \ldots K - 1$) as the interval since a new node starts transmitting, unsynchronized, until it is integrated in the team and starts transmitting in its own slot (A–D in Fig. 6) and we refer to it as $T_{i,join_K}$. In the absence of errors, the agreement phase takes one round (A–B). However, if this phase is affected by errors, it is extended another round, and possibly more rounds until it succeeds.

The round reconfiguration phase starts with the new node already transmitting with status *Running*. This phase can be further divided in two intervals, until the next reference packet is received (B–C) and from then on until the first transmission of the node in its slot in the new round configuration (C–D).

The first of these intervals, i.e., until the reference packet is received, depends on the relative phase between the new node initial transmissions and the initial round configuration. In the best case, it lasts the transmission time of the reference packet t_{packet} and in the worst-case, it takes one round plus the largest delay Δ_{K-1} that the reference packet might have suffered in this round (still using the initial configuration). In the presence of packet losses, this interval can be further extended by one or more integer rounds.

The second interval is given by the offset of the new node slot, i.e., $i \times T_{xwin_K}$ where i is the dynamic ID of the new node in the new round configuration. Therefore, $T_{i,join_K}$ can be bounded by Eq. 6, without considering packet losses.

$$T_{tup} + t_{packet} + i \times T_{xwin_K} \leq T_{i,join_K} \leq 2 \times T_{tup} + \Delta_{K-1} + i \times T_{xwin_K} \tag{6}$$

When considering the impact of errors and consequent packet losses, the upper bound needs to include the extra rounds incurred as given by Eq. 7 where $n(b, p)$ is the number of extra rounds that need to be considered for a given probability p of successful packet reception and given a bit error rate b.

$$T_{i,join_K} \leq (2 + n(b,p)) \times T_{tup} + \Delta_{K-1} + i \times T_{xwin_K} \tag{7}$$

Using Eq. 4 we can upper bound the last two terms to $1.5T_{tup}$ independently of ε, i and K, or to T_{tup} as long as $\varepsilon \leq 1 - 1/K$, which is a frequent configuration. In this latter case, the joining process can be upper bounded by T_{join} in Eq. 8.

$$T_{join} \leq (3 + n(b,p)) \times T_{tup} \tag{8}$$

Given its complexity and dependency on the actual error patterns, we use an experimental characterization of the $n(b, p)$ function.

5.3 Performance Assessment

In this section we show experimental results that allow assessing the performance of the Reconfigurable and Adaptive TDMA protocol, both in laboratory environment as well as in actual operation.

The first experiment compares the protocol against a common TDMA approach with a rigid slot structure implemented with clock synchronization using Chrony.[1] In this experiment there is a residual load of uncontrolled external background

[1] http://chrony.tuxfamily.org.

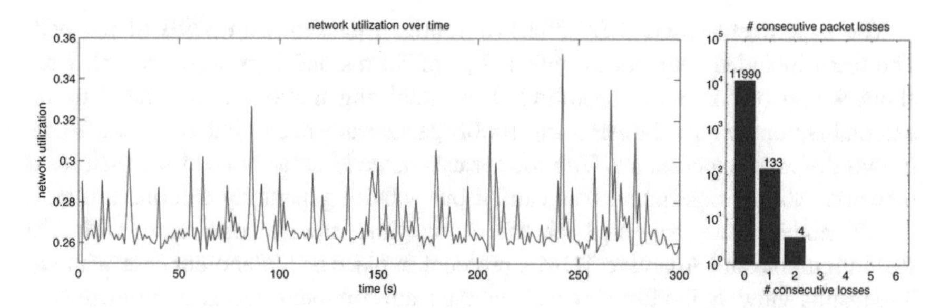

Fig. 7 TDMA with rigid slot structure using clock synchronization (instantaneous network utilization and histogram of consecutive packet losses)

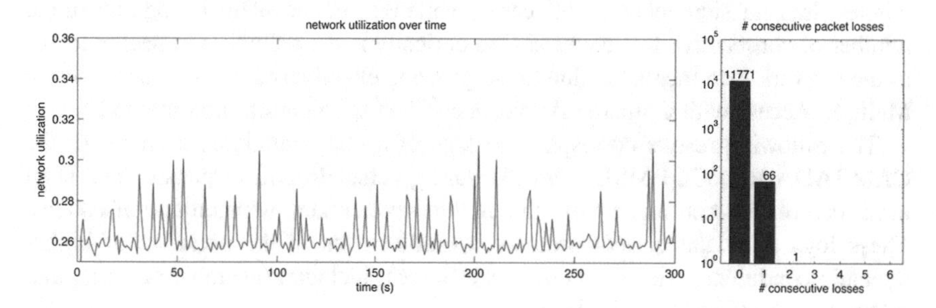

Fig. 8 TDMA with adaptive slot structure (instantaneous network utilization and histogram of consecutive packet losses)

traffic plus a periodic interference caused by an external node issuing a `ping` command to the Access Point (AP) with 1 kB every 5 ms. The TDMA round was configured with a period 99.5 ms and with four slots.

Figure 7 shows the results with Chrony. The rigid periods of the TDMA framework and `ping` packets create a sliding phase that causes a periodic critical interference approximately every 20 s with the same slot but with different offsets for different slots. Several such intervals are clearly visible in the spikes that affect the network load. On the right we can see a histogram of successful team transmissions, single lost packets and multiple consecutive lost packets. Overall, we observed 1.17 % losses.

On the other hand, Fig. 8 shows the same case but using the Reconfigurable and Adaptive TDMA with T_{tup} = 99.5 ms. The network load plot now shows a more irregular pattern, as expected due to the phase adaptation of the TDMA round which average duration increased to 100.4 ms. In terms of losses, we observed a reduction to 0.56 % and without affecting consecutive packets.

Another relevant comparison is with non-synchronized periodic transmissions as if, for example, the agents in the team were transmitting independently. In this case, differences in the clocks lead to phase drifts and occasionally to several agents transmitting at the same time creating critical interference periods.

The experiments used IEEE 802.11b multicast packets with 379B of payload. The team included four nodes with a T_{tup} of 50 ms and logs were extracted for about 9 min of continued operation. The interfering traffic was generated by an external laptop *pinging* the AP using 1000B packets at a rate of one per 5 and 10 ms in two different experiments. The former case already corresponded to a saturated network. A third experiment was carried out without generating external traffic.

All nodes were triggered at the same time but in one case with the Reconfigurable and Adaptive TDMA protocol in place and in another case without. The results show, in the former case, that the team communications are immediately reorganized in the synchronized TDMA framework, reducing collisions, while in the latter the agents will continue colliding, generating a period of poor channel quality, independently of the channel load. Figure 9 shows the histograms of the wireless transmission delay in both cases, while Fig. 10 shows the histograms of the number of consecutive lost packets. One curiosity is the reduction in packet losses as the network load increases, due to the growing effectiveness of the Carrier Sense Multiple Access with Collision Avoidance (CSMA/CA) arbitration method.

The following results correspond to logs of a real operational scenario of the CAMBADA RoboCup MSL team [3] during actual RoboCup games. The team used six mobile robots and a base station, thus seven nodes, with a round of 100 ms. These logs show the correct protocol operation even in the presence of highly dynamic conditions, with frequent packet losses, including asymmetric ones, and robots that leave and join the team.

Figure 11 shows the round structure (top) and the cardinality of the membership (below), i.e., the number of active agents in the team. Each line in the round structure represents the offset of a slot transmission with respect to the beginning of the round, i.e., transmission of the reference in slot 0. It illustrates the maximization of the time interval between consecutive transmissions of team members as the team composition changes (distance between consecutive lines).

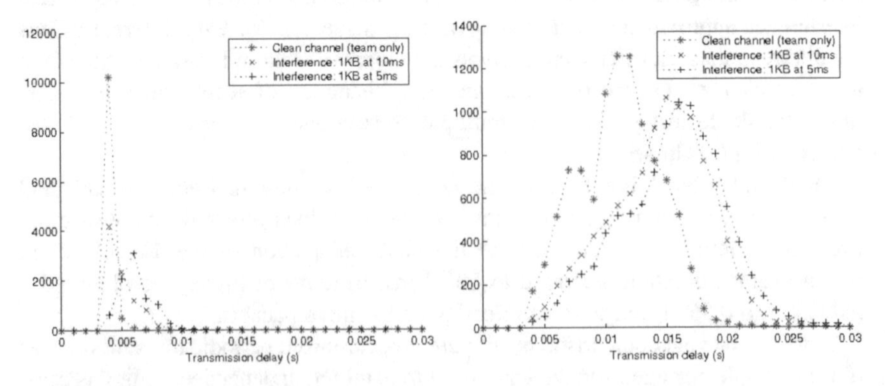

Fig. 9 Histograms of the transmission delay with reconfigurable and adaptive TDMA (*left*) and without synchronization (*right*)

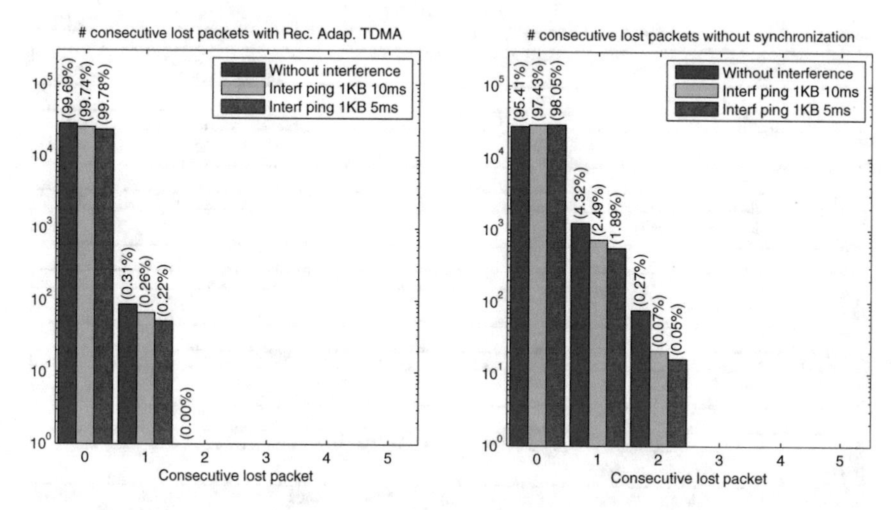

Fig. 10 Histograms of the number of consecutive lost packets with reconfigurable and adaptive TDMA (*left*) and without synchronization (*right*)

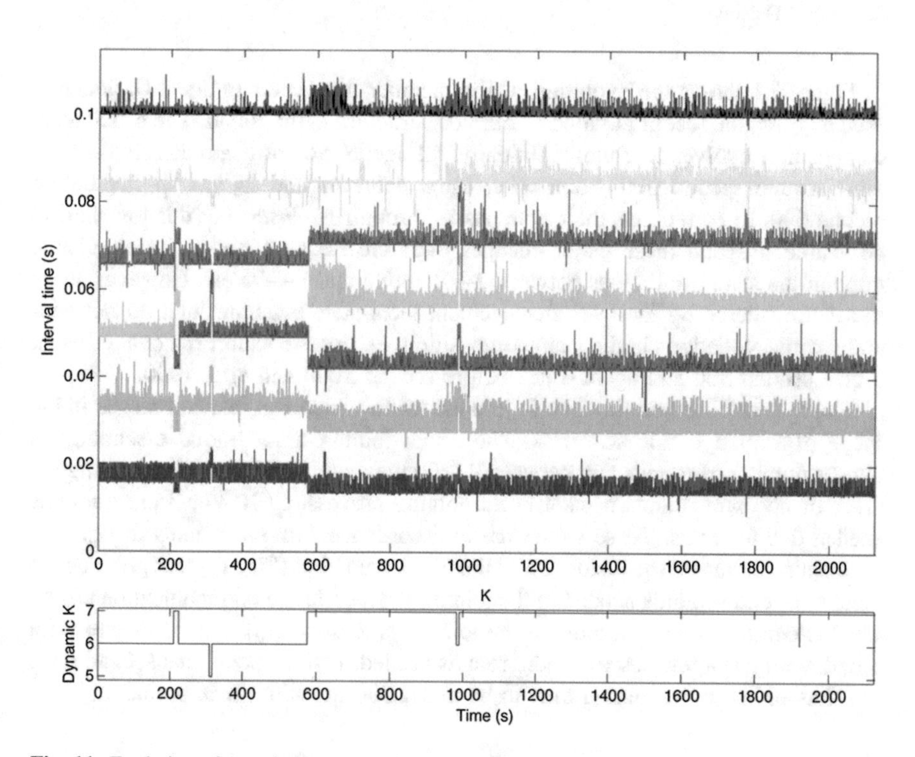

Fig. 11 Evolution of round structure and number of nodes in the team

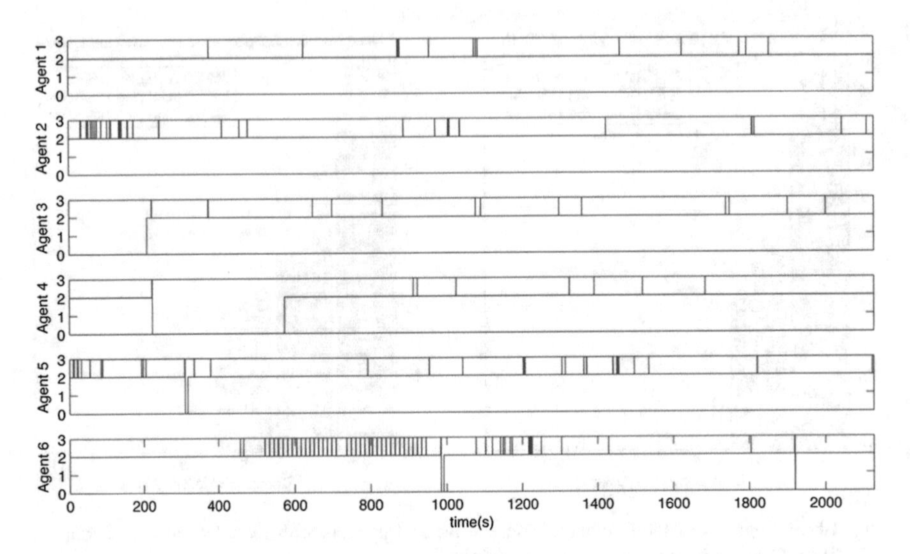

Fig. 12 Membership vector dynamics in agent 0 (agents states code: *0* Not Running; *1* Insert; *2* Running; *3* Delete)

Figure 12 shows the evolution of the membership vector in agent 0, which is used to guide the reconfiguration of the slots structure in the round. The actual team composition evolves as follows. Initially, all agents except 3 are active. Agent 3 joins around second 200. Soon after, agent 4 leaves and rejoins shortly before second 600. In both cases, their state passes through 1—*Insert* but it is too short to be visible. Beyond these major changes there are frequent packet losses, visible through the state oscillations between 2—*Running* and 3—*Delete*. However, these losses are filtered by the team management local state machines and do not generally cause variations in the team composition, except for a short removal of agent 5 near second 300 and agent 6 just before second 1000 and after 1900.

Finally, we observed the time to join the team, which is a relevant metric in the scope of very dynamic scenarios. The actual joining times varied essentially in integer number of rounds and between 2 (200 ms) and 5 (500 ms), representing the cases of no extra rounds needed in the joining process, or 1, 2 or 3 extra rounds needed due to errors. These values are in accordance with the bounds in Eq. 7.

We also extracted the histogram of the $n(b, p)$ function (Fig. 13) that gives us the number of extra rounds needed in the joining process due to communication errors. The histogram shows that most of the joining processes require none or one extra round, with just a few cases of more rounds needed. In this operational scenario, we can thus estimate the joining time to be under 500 ms with 96 % probability.

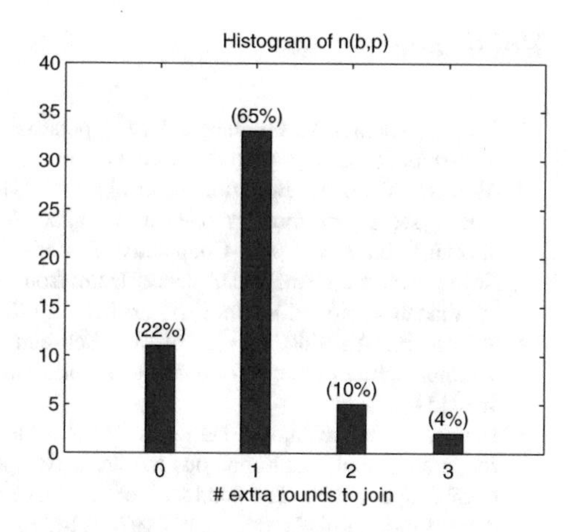

Fig. 13 Distribution of extra rounds needed in a joining process due to errors

6 Conclusion

This chapter addressed the problem of developing applications for Mobile Cyber-Physical Systems (M-CPS) composed of cooperative mobile autonomous agents. It focused on the communications and middleware platform on which most cooperative behaviors rely. For such purpose, we developed a platform that allows sharing state among team members while abstracting away communication. It relies on a shared memory middleware that extends the traditional Blackboard concept with local data proxies that also include data age information. This middleware is then connected to a TDMA protocol that updates the data proxies and supports dynamic addition and removal of team members. We named these components the Real-Time Database (RTDB) and the Reconfigurable and Adaptive TDMA protocol. We briefly explained their main features and we then showed several experimental results that confirm, in particular, the properties of the communication protocol, which operates over IEEE 802.11 in infra-structured mode.[2]

This platform has been validated in practice in demanding operational scenarios such as robotic soccer in the RoboCup Middle Size League. The communications protocol was also extended to mesh networks that give a higher flexibility in topology control. Currently the platform is being reworked to suit vehicular applications, which require, among other features, relative localization services to support applications like platooning and smart intersections.

[2]Code available at www.bitbucket.org/fredericosantos/rtdb/.

References

1. Cao, Y., Kahng, A., Fukunaga, A.: Cooperative mobile robotics: antecedents and directions. In: Arkin, R., Bekey, G. (eds.) Robots colonies (1997). doi:10.1007/978-1-4757-6451-2_1
2. Akin, H., Birk, A., Bonarini, A. et al.: Two "Hot Issues" in cooperative robotics: network robot systems, and formal models and methods for cooperation. In: Technical report, EURON Special Interest Group on Cooperative Robotics (2008)
3. Neves, A. et al.: CAMBADA soccer team: from robot architecture to multiagent coordination. In: Vladan Papić (ed.) Robot soccer. InTech (2010). doi:10.5772/7353
4. Santos, F., Almeida, L., Lopes, L.: Self-configuration of an Adaptive TDMA wireless communication protocol for teams of mobile robots. ETFA (2008). doi:10.1109/ETFA.2008.4638554
5. Santos, F., Almeida, L., Pedreiras, P.: A real-time distributed software infrastructure for cooperating mobile autonomous robots. ICAR (2009). Munich
6. Kernbach, S., et al.: The handbook of collective robotics: fundamentals and challenges. CRC Press, Florida (2013). doi:10.4032/9789814364119
7. Urcola, L., Montano, L.: Cooperative robot team navigation strategies based on an environment model. IROS (2009). doi:10.1109/IROS.2009.5354243
8. Hoff, N., Sagoff, A., Wood, R. et al.: Two foraging algorithms for robot swarms using only local communication. ROBIO (2010). doi:10.1109/ROBIO.2010.5723314
9. Ranjan, A.: Underwater wireless communication network. Adv. Electron. Electr. Eng. J. **3**(1), 41–46 (2013)
10. Tardioli, D., Mosteo, A., Riazuelo, L. et al.: Enforcing network connectivity in robot team missions. Int. J. Robot. Res. (2010) doi:10.1177/0278364909358274
11. Mora, F.: Bring an atomic clock to your home with chrony. Linux J. **101** (2002)
12. Lamport, L.: Time, clocks, and the ordering of events in a distributed system. communications of the ACM **21**(7) (1978). doi:10.1145/359545.359563
13. Oliveira, L., Li, H., Almeida, L. et al.: RSSI-based relative localisation for mobile robots. Ad Hoc Networks (Elsevier) (2014). doi:10.1016/j.adhoc.2013.07.007
14. Haghani, E., Krishnan, M., Zakhor, A.: Adaptive carrier-sensing for throughput improvement in IEEE 802.11 Networks. GLOBECOM (2010). doi:10.1109/GLOCOM.2010.5683231
15. Balador, A., Movaghar, A.: The novel contention window control scheme for IEEE802.11 MAC protocol. NSWCTC (2010). doi:10.1109/NSWCTC.2010.167
16. Bartolomeu, P., Fonseca, J., Vasques, F.: Implementing the wireless FTT protocol: a feasibility analysis. ETFA (2010). doi:10.1109/ETFA.2010.5641350
17. Costa, R., Portugal, P., Vasques, F. et al.: A TDMA-based mechanism for real-time communication in IEEE 802.11e networks. ETFA (2010). doi:10.1109/ETFA.2010.5641340
18. Santos, F., Almeida, L., Lopes, L. et al.: Communicating among robots in the RoboCup middle-size league. In: Baltes, J., Lagoudakis, M., Naruse, T. et al. (ed.) RoboCup 2009: robot soccer world cup, LNCS 5949 (2010). doi:10.1007/978-3-642-11876-0_28
19. Fortino, G., Guerrieri, A., Russo, W., Savaglio, C.: Middlewares for smart objects and smart environments: overview and comparison. In: Fortino, G., Trunfio, P. (eds.) Internet of things based on smart objects: technology, middleware and applications. Springer, pp. 1–27 (2014). doi:10.1007/978-3-319-00491-4_1
20. Mohamed, N., Al-Jaroodi, J., Jawhar, I.: Middleware for robotics: a survey. RAM'08 (2008). doi:10.1109/RAMECH.2008.4681485
21. Mohamed, N., Al-Jaroodi, J., Jawhar, I.: A review of middleware for networked robots. Int. J. Comput. Sci. Netw. Secur. **9**(5), 139–148 (2009)
22. Corkill, D.: Collaborating software: blackboard and multi-agent systems & the future. In: International lisp conference. New York, USA (2003)
23. Hayes-Roth, B.: A blackboard architecture for control. Artif. Intell. **26**(3), 251–321 (1985). doi:10.1016/0004-3702(85)90063-3

ANIMO, Framework to Simplify the Real-Time Distributed Communication

Yamnia Rodríguez, Carlos Alejo, Irene Alejo and Antidio Viguria

Abstract This paper presents a communication framework developed for interconnecting multi-systems based on the Data Distribution Service (DDS). The newly built framework, called ANIMO, facilitates the integration of DDS in an application and the interoperability between the different data types of the Cooperating Objects (COs) with the great feature of real-time. Furthermore, a powerful tool to generate code has been developed bringing the incorporation and the updating of the data types as a very easy and simple task. In addition, a novel module has been performed to give the capacity to communicate the ANIMO framework with the ROS middleware that makes even easier the integration of mobile robots as another Cooperating Object. This paper explains the complete architecture of the ANIMO framework, its diversity of possibilities and two principal works where it has been applied. One is a distributed simulator to validate embedded control algorithms. The other is the task of the supervision and message passing between quadrotors in an experiment of coordination and cooperation involving multiple aerial vehicles.

Keywords Software communication layer · Real-time · Distributed · Multi-systems · Quality of service · Interoperability

Y. Rodríguez (✉) · C. Alejo · I. Alejo · A. Viguria
Centre for Advanced Aerospace Technologies (CATEC),
41309 La Rinconada (Sevilla), Spain
e-mail: yrodriguez@catec.aero

C. Alejo
e-mail: calejo@catec.aero

I. Alejo
e-mail: ialejo@catec.aero

A. Viguria
e-mail: aviguria@catec.aero

A. Guerrieri et al. (eds.), *Management of Cyber Physical Objects in the Future Internet of Things*, Internet of Things,
DOI 10.1007/978-3-319-26869-9_4

1 Introduction

The increasing usage of heterogeneous systems jointly, such as variety of sensors, microcontrollers, PCs, robots, etc., all of which scattered in an environment, implicates a greater effort on inevitable tasks of integration and communication between them. In these cases, where the importance of the work lies in the development of applications and algorithms and not how the system data are available and interchangeable, a communication software with fast tuning capabilities and high performance is desired. Also, many engineering applications related to smart cities require this data fusion to be performed in real time. For instance, efficient distributed video surveillance requires that real-time constraints be respected or, at least, quality of service guarantees (QoS) be provided [1]. In order to fulfil this necessity, a communication framework, called ANIMO, has been developed. Along this paper, the ANIMO framework will be explained.

The communication model underlying in network software is the most important factor in how applications communicate. The communications model impacts the performance, the ease to accomplish different communication transactions, the nature of detecting errors, and the robustness to different error conditions. Unfortunately, different communication models are better suited to handle different classes of application domains. The three main types of network communication models are: point-to-point, client-server and publish-subscribe.

For many applications related to cooperating objects in smart cities, the publish-subscribe communication model is the one that best suits its requirements since it is the best choice for systems with complex time-critical data flows [2]. Computer applications (nodes) "subscribe" to data they need and "publish" data they want to share. Messages pass directly between the publisher and the subscribers, rather than moving into and out of a centralized server. Publish-subscribe communication architectures are good for distributing large quantities of time-sensitive information efficiently, even in the presence of unreliable delivery mechanisms.

Taking into account the above, distributed applications and satisfying real-time, the ANIMO framework has been developed based on the open standard of the Data Distribution Service for Real-Time Systems (DDS) from the Object Management Group (OMG). A well known open-source implementation of the DDS is the RTI (Real Time Innovations) DDS, which has been used in this work. Studies show that DDS implementations perform significantly better than non-DDS alternatives and are well-suited for certain classes of data-critical DRE information management systems [3].

In addition, other software modules have been built facilitating the integration of systems and devices, the interoperability between their different data types and the maintenance and enlargement of the framework itself. Moreover, the network features needed for an application can be adjusted to the network constraints through a wide Quality of Service controls (configuration parameters), achieving a high throughput.

The paper is organized as follows. In the next section the newly developed framework, called ANIMO, based on the DDS is described. In Sect. 3, a brief description of two research projects, where the ANIMO framework has been successfully applied, is exposed and illustrated with results obtained from experiments. In Sect. 4, some results about performance tests done using the ANIMO framework are presented. Finally, the last section is devoted to final conclusions.

2 The ANIMO Framework: Aiding to the Systems Intercommunication

As commented, the ANIMO framework has been developed to make easier the integration between different heterogeneous systems such as sensors, robots, systems, etc. that can be part of any smart cities application. It has a scalable architecture that is formed by the following modules:

- The communication layer (CL). This is the framework core based on DDS.
- The hardware abstraction layers (HAL). There is a HAL for each device or Cooperating Object to be incorporated into the framework. This layer is responsible for abstracting the peculiarities of the device and converting the data types used by its driver to the types defined in ANIMO. The different HALs are libraries that wrap the device drivers to integrate them easily.
- The services of devices as high level applications (HLAs), that using the communication layer and the hardware abstraction layer, send the device data and receive the control command by DDS. There are already several devices integrated in ANIMO such as the haptics, the cyber gloves, the IS900 tracker, the webcams and the testbed accessing to the quadrotors. In this way, the devices are available for the possible HLAs. It is shown in Fig. 1.
- The ROS Bridge module gives the capacity to communicate the ANIMO framework with the ROS middleware[1] that makes even easier the integration of mobile robots as another Cooperating Object, i.e. the ROS world is accessible in ANIMO. This characteristic could be very interesting for smart cities applications where it is needed to interconnect robots with other smart sensors.
- A powerful tool to generate code brings creating ANIMO extensions as a very easy and simple task.

The benefits that the ANIMO framework provides to the applications are numerous. For example it accelerates development since it replaces low-level communications code with high-level interfaces. Also, it combines asynchronous

[1]ROS (Robot Operating System) provides libraries and tools to help software developers create robot applications. It provides hardware abstraction, device drivers, libraries, visualizers, message-passing, package management, and more. ROS is licensed under an open source, BSD license [4].

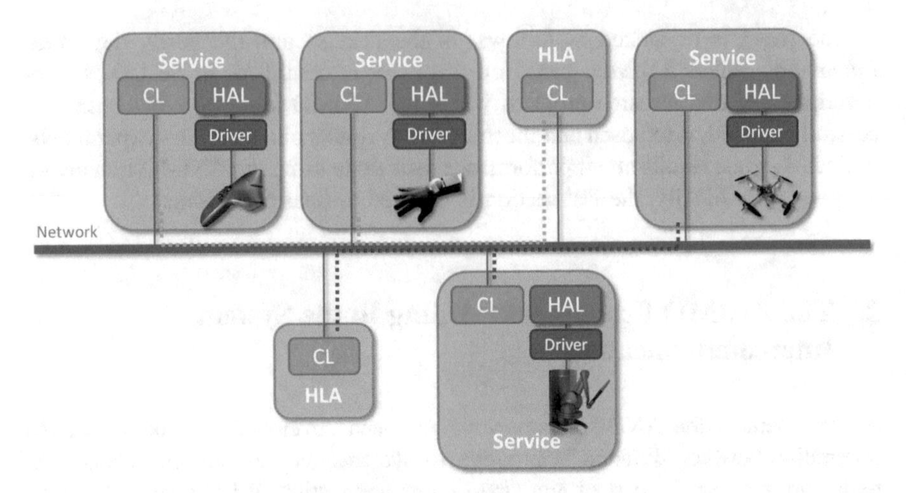

Fig. 1 Example of system intercommunication using the ANIMO framework

publish/subscribe messaging and real-time data management, it automatically discovers and routes data between publishers (writers) and subscribers (readers) to the same topic, eliminates dependence on start-up order, automatically synchronizes state if there are disconnections or/and reconnections, and a change of the QoS used in the application means only a change in the configuration file avoiding the recompilation of the application.

2.1 Architecture of the Communication Layer

Given that the DDS API is very extensive and it is desired that the communication layer be reusable and simple, hiding the complexity of DDS, the ANIMO framework has built its own communication layer which can be integrated into any application requiring distributed software. Its architecture is detailed in Fig. 2.

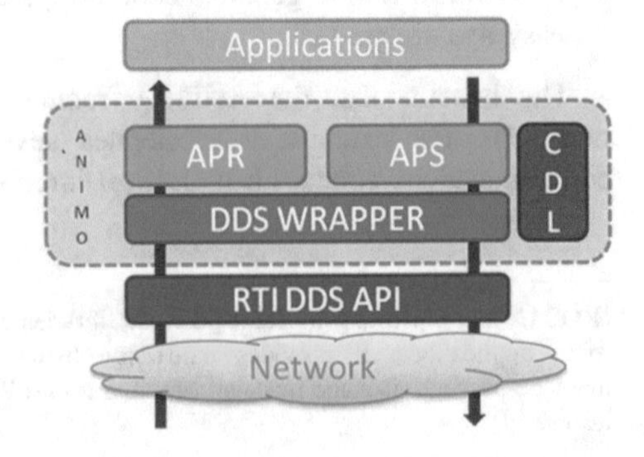

Fig. 2 The ANIMO communication layer architecture

The communication layer is implemented in the C++ programming language and it is available for Linux and Windows platforms as dynamic library.

The ANIMO framework has a first layer of DDS abstraction (DDS wrapper). It creates the topics, the participants, the publishers, the data writers, the subscribers and the readers. However, in order to further facilitate the use of the framework, ANIMO integrates another layer of abstraction that provides access points for sending (APS) and receiving (APR) data to the applications reducing the number of objects to create.

The common data types or structures that are exchanged between the applications are defined in Common Data Library (CDL), which gives support to the communication layer and the applications. At present, the CDL has more than forty different data types such as position, rotation, velocity, angular speed, flight plan, MAVLink messages, etc. Thanks to the powerful tool built to generate code, the incorporation and the updating of the data types is a very easy and simple task.

2.2 A Rapid and Easy Way to Extend the ANIMO Framework

A feature that adds more value to the framework is to have a code generator for updating the framework. The code generator is joined to a useful GUI program that helps the developer in the maintenance task. From a description file of the data types that will travel through the network, all the ANIMO framework code that depends on the data types is self-generated. Then, these files are copied to the respective projects and compiled. Finally, the installation packages are built to reinstall them updating the framework libraries on the computer.

XML has been chosen as the description file format because it is well-known by most developers. Managing the information that contains, it is possible to auto generate the ROS messages and actions for the ROS packages and ROS Bridge module, and the IDLs and the C++ code for the ANIMO framework core. The code generator is based on the Apache Velocity Project [5].

2.3 Connecting with the ROS Middleware

A novel module, called ROS Bridge, has been developed to give the capacity to communicate the ANIMO framework with the ROS middleware that makes even easier the integration of mobile robots as another Cooperating Object. More specifically, ROS Bridge allows the automatic generation of a data parser between the ANIMO framework and ROS using a XML configuration file. In Fig. 3 it can be observed the link of the ANIMO framework core with the ROS Bridge module. The ROS Bridge API wraps the ROS API providing methods for the publication

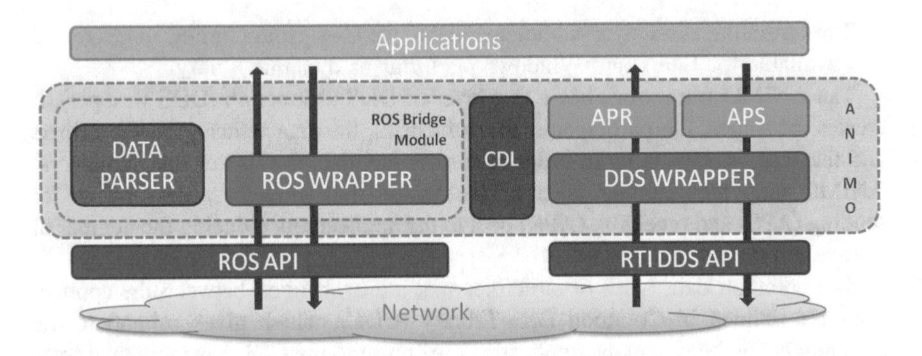

Fig. 3 The ROS Bridge module architecture

and the subscription of the ROS messages. Moreover, separate ROS networks can be communicated by a DDS network using the ROS Bridge module. In this way, the isolated clients and servers of the ROS actions achieve to be connected and are able to interchange data across the ANIMO framework.

3 Example of Application Projects

The following sections explain two principal works where ANIMO has been applied. One is a distributed simulator to validate real embedded control algorithms. The simulator has been developed in the European Commission's 7th Framework Programme (FP7) project, EC-SAFEMOBIL [6]. The other is the task of the supervision and message passing between quadrotors in an experiment of coordination and cooperation involving multiple aerial vehicles. This task is still developing in the ARCAS project [8]; it is also a FP7 project.

Following are some other noteworthy works where ANIMO is involved:

- Real-time visualization of the experiments being carried out in the testbed of FADA-CATEC through a scaled, virtual world, with the possibility of 3D projection.
- Manipulation of an articulated arm of two degrees of freedom using a haptic device.
- Simulator for the formation of operators of Remotely Piloted Aircraft Systems (RPAS), implicating communication between the Ground Control Station (GCS), the simulation engine, the simulated aircraft with real embedded control algorithms, the instructor station and the databases. This project is still being developed.
- Human Machine Interface (HMI) for the navigation in a 3D virtual world (zoom, translation and rotation) commanded from the gestures collected using a cyber glove and the IS900 tracker.

3.1 Distributed Simulator to Validate Embedded Control Algorithms

The EC-SAFEMOBIL project is devoted to the development of sufficiently accurate common motion estimation and control methods and technologies in order to reach levels of reliability and safety to facilitate unmanned vehicle deployment in a broad range of applications [6]. With this goal, several scenarios are defined in the project to be performed both in simulation and in real experiments. The distributed simulator offers the opportunity to test the developed methods in the project in a repeatable and thorough manner, validating embedded control algorithms before executing them in a real experiment, regarding scalability, fault-tolerance, trajectory planning and cooperative tracking. The most important scenarios are:

- Rotary-wing UAV landing on a ship (see Fig. 4a).
- Autonomous distributed warehousing traffic and control (see Fig. 4b).
- Tracking for surveillance. Many UAVs (Unmanned Aerial Vehicles) follow their targets (see Fig. 4c).

The simulator has been developed based on the VT MÄK tools [7], VR-Forces (back-end module or simulation engine) and VR-Vantage (front-end module or

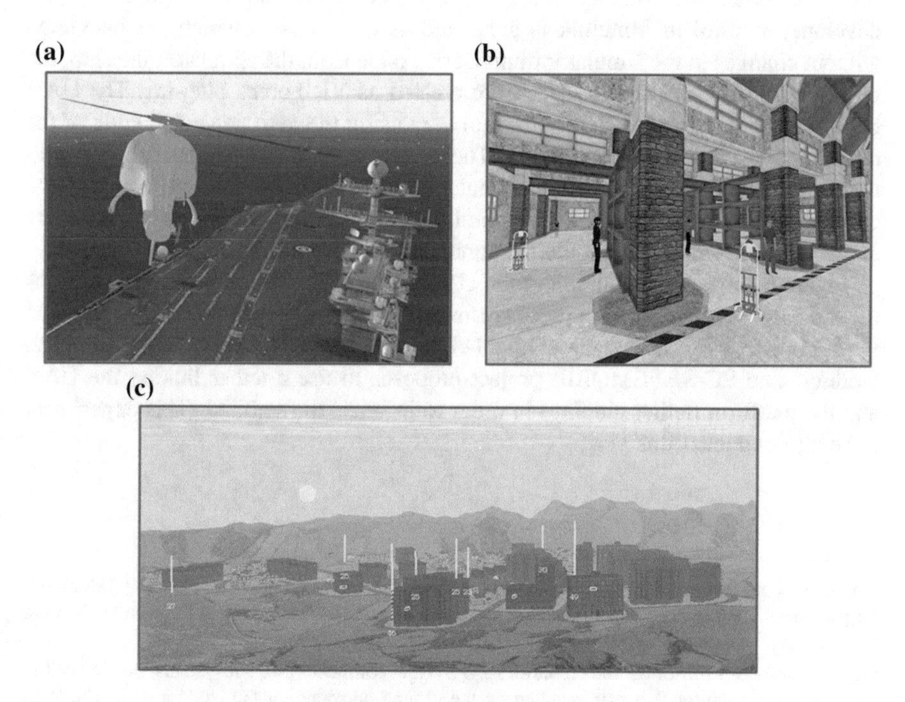

Fig. 4 Images of the graphic interface of the simulator

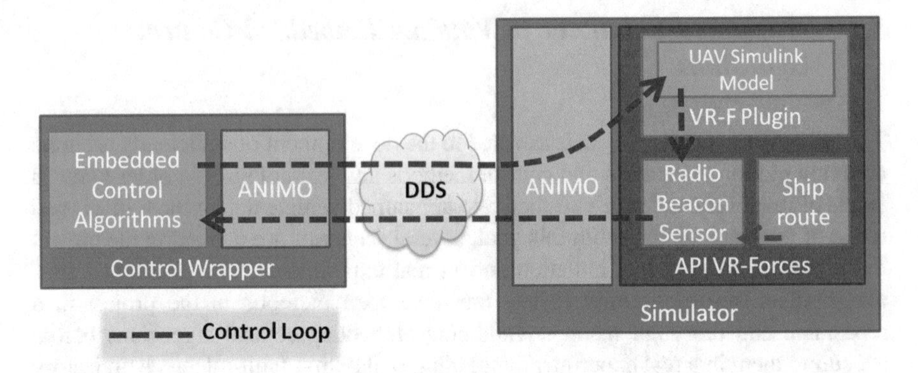

Fig. 5 Main components of the distributed simulator in the UAV landing on a ship scenario

visual image generator), and has the great feature that the simulations can be visualized in 3D. The rotary-wing UAV landing on a ship scenario will be explained as example. The modules implicated in this scenario constituting the simulator framework are shown in Fig. 5. There are two programs exchanging data by using the ANIMO framework core.[2] One is the simulation engine; it is responsible for the synchronization between all modules managing the simulation time. The other is a wrapper of the real UAV control embedded algorithms. The developed control in Simulink is generated as C++ classes which are integrated without changes in the simulator framework. Aside from the simulator, the ship, the radio beacon sensor[3] and the UAV are created as VR-Forces plug-ins. The UAV behaviour is simulated in a very truthful way using the auto generated code of the developed UAV model in Simulink. The control loop is implemented in the simulator as shown in Fig. 5. In each simulation step, the RBS data are inputs in the control algorithms which act in the simulated UAV. After that, the new UAV state along with the ship state update the simulated sensor closing the loop.

Figure 6 shows the evolution of the UAV and target position in wave conditions (a) and calm sea (b). When the sea is calm, the UAV is able to land. However, when the sea is wavy, it is necessary another helping mechanism to carry out the complete landing. The EC-SAFEMOBIL project proposes to use a tether linking the UAV and the platform (tether guiding) in order to increase the stability. This experiment is being conducted this year.

[2]In the tracking for surveillance scenario, can be observed in the Fig. 4c, there are many UAVs and targets, therefore, the simulator framework is composed of a wrapper program for each UAV more the simulation engine program.

[3]It is the proposed sensor for the scenario. RBS System comprises two subsystems, one located in the UAV itself and another one installed on the ground surrounding the landing area. The RBS airborne system provides a relative position to the UAV on board computer (relative to the moving landing pad, located on the ship, in body fixed frame).

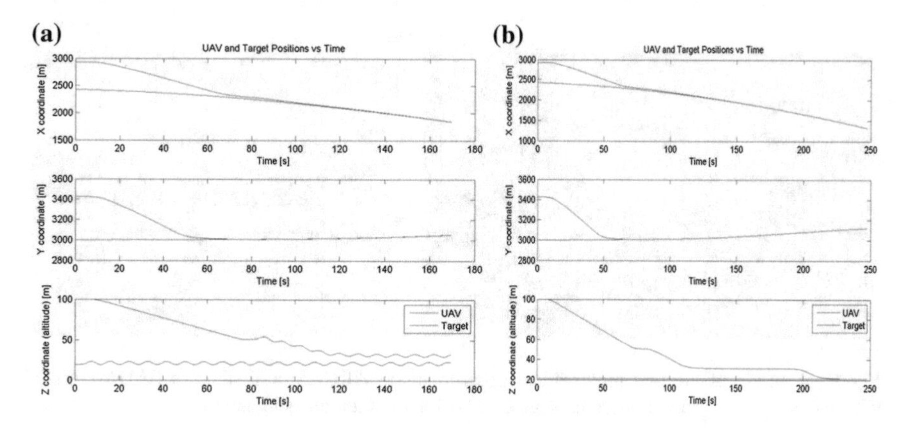

Fig. 6 The evolution of the UAV and target position in wave conditions (**a**) and calm sea (**b**) in the UAV landing on a ship scenario

3.2 Experiment of Coordination and Cooperation Involving Multiple Aerial Vehicles

The ARCAS project proposes the development and experimental validation of the first cooperative free-flying robot system for assembly and structure construction [8]. For the missions involving inspection tasks and translation of structures, experiments of coordination and cooperation of multiple aerial vehicles and collision avoidance are proposed. At present, these experiments have been executed in simulation using Gazebo (see Fig. 7b). The real experiments are programmed for early March this year using the testbed[4] in FADA-CATEC (see Fig. 7a).

The architecture of the multivehicle experiments is presented in Fig. 8. The modules executed on the ground station, the Trajectory Planning (TP) and the Global Supervisor (GS) are connected to a DDS network. Whereas the modules executed on board, the Trajectory Generation (TG), the Collision Avoidance (CA) and the UAV Abstraction Layer (UAL) are connected by a ROS network. The two networks are united thanks to the ROS Bridge (ROSB) forwarding the ANIMO topics to the ROS topic and vice versa. The ROSB runs on the ground station. In the beginning, the GP sends some waypoints of the desired trajectory of all involved UAVs to the TP. Then the TP, which knows the world, plans the trajectories for each UAV avoiding the obstacles of the area and send the trajectories back to the UAVs. The trajectories arrive to the UAVs through the ROS Bridge module. At that

[4]This testbed is based on an indoor positioning system that uses 20 VICON cameras. This system can calculate the position and attitude of any moving object within the volume of the testbed (15 × 15 × 5 m) in real time (with an update rate of up to 500 Hz).

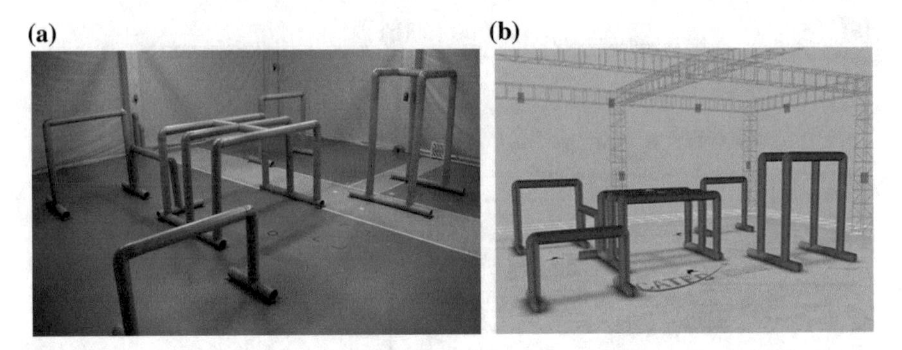

Fig. 7 Scenario of the first multivehicle experiments in ARCAS. **a** Testbed of FADA-CATEC with mock-up. **b** Virtual environment in Gazebo for the previous simulations

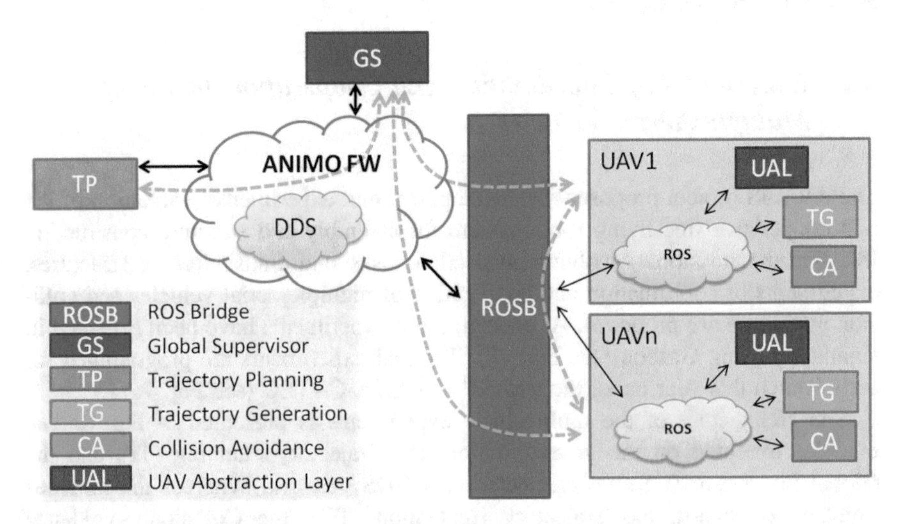

Fig. 8 Architecture of the experiment of coordination and cooperation involving multiple aerial vehicles

point, the TG generates the control references for its UAV. These control references act in the low-level software using UAL. Furthermore, the control references are modified from CA if a collision with other UAVs is detected. To accomplish this it is necessary that each UAV know the state of every other UAV, which is achieved using the ROS Bridge module.

Figure 9 shows the results of an experiment with three UAVs. Figure 9a is an experiment without executing the CA module. Therefore, there is a collision implicating two UAVs which go to ground. In Fig. 9b it can be observed that the collision is avoided because the CA module is being executed.

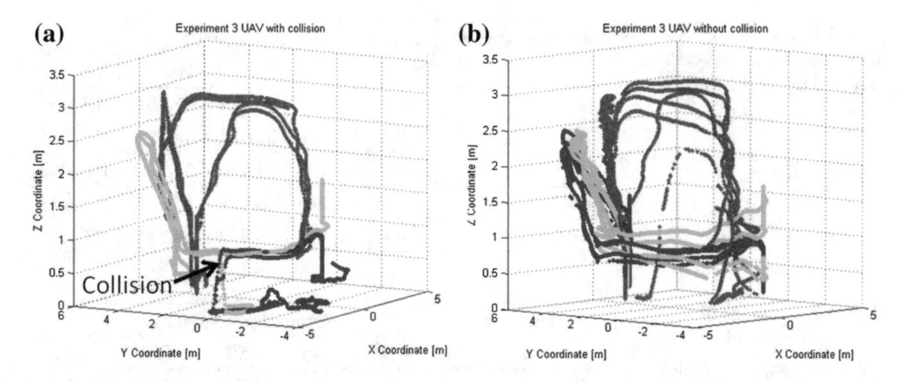

Fig. 9 Results of an experiment with three UAVs. **a** Without executing the CA module. **b** Executing the CA module

4 ANIMO Performance Tests

The performance tests presented in this chapter were carried out as part of the works regarding the EC-SAFEMOBIL project, which had the ANIMO framework integrated. After an initial integration, the middleware was working with the different scenarios, but further actions were required in order to improve the middleware performance and reach the necessary level to use it in the real experimentation progress. During 2014, these improvements were implemented and tested. This set of tests allowed us to obtain a measure of the middleware capability and to decide if its nominal behaviour is sufficient to be further used in the scenarios.

The performance tests were performed in FADA-CATEC's testbed, in order to facilitate the measurement of the required metrics, with two helper machines at another facility. The laptops used for this were two identical IBM ThinkPad T60 laptops with Windows 8.1 as operating system. The network connectivity was established using a wired Gigabit LAN.

4.1 Payload Tests

The idea of these tests was to analyse the influence of different payload sizes on the transmission behaviour of the EC-SAFEMOBIL communication middleware. Messages were sent for 10 min, containing a payload between 1 byte and 10,000 bytes, and with a sending rate ranging between 10 messages per second and 200 messages per second.

Table 1 shows the test results taking into account the message payload and the sending rate. The table contains the exact amount of messages sent through the network during the tests. The results show, that as expected, a higher payload size reduces the number of messages that can be sent in the same amount of time, since

Table 1 Number of received messages with a specific payload using the middleware

Payload (bytes)	Rate (messages per second)				
	10	25	50	100	200
1	5997	14,969	29,484	57,791	108,906
64	5997	14,971	29,481	58,125	107,376
100	5997	14,973	29,436	57,780	108,942
6400	5967	14,795	28,972	55,225	101,672
10,000	5963	14,775	28,822	54,585	100,251

Table 2 Number of messages lost (with a specific payload) using the middleware

Payload (bytes)	Rate (messages per second)				
	10	25	50	100	200
1	0	0	0	0	0
64	0	0	0	0	0
100	0	0	0	0	0
6400	0	0	0	0	1
10,000	0	0	0	2	60

Table 3 Percentage obtained comparing the real results with the ideal throughput

Payload (bytes)	Rate (messages per second)				
	10	25	50	100	200
1	99.95	99.79	98.28	96.32	90.76
64	99.95	99.81	98.27	96.88	89.48
100	99.95	99.82	98.12	96.30	90.79
6400	99.45	98.63	96.57	92.04	84.73
10,000	99.38	98.50	96.07	90.98	83.54

more time is needed to physically transmit the information through the network. Table 2 pictures the message loss during the experiments caused by the unreliable but fast UDP protocol. As it is shown, message loss only occurs when sending with a very high rate or with a very large payload.

Table 3 shows the percentage of the throughput achieved in the experiments compared to the theoretically ideal maximum throughput. The ideal maximum throughput is the theoretical number of messages sent in the 10 min of the experiments duration. In most cases the middleware performs with values greater than 90 % of the theoretical maximum. And even in the worst tested scenarios the achieved 83 % is higher than the required 75 % maximum throughput, as set in the requirements of the project.

4.2 *Arrival Time Tests*

This experiment shows how long it takes to send a pre-defined number of messages with a payload of 64 bytes. For this, five different amounts of messages have been

Table 4 Time spent (in seconds) sending all the required messages using the middleware

Number of messages	Rate (messages per second)			
	25	50	100	200
99	3.932	1.979	1.031	0.537
499	19.963	10.131	5.188	2.738
999	39.996	20.349	10.348	5.500
4999	200.524	104.541	55.223	27.613
9999	401.750	208.950	109.111	60.043

Table 5 Time spent (in seconds) sending all the required messages in ideal conditions

Number of messages	Rate (messages per second)			
	25	50	100	200
99	3.960	1.980	0.990	0.495
499	19.960	9.980	4.990	2.495
999	39.960	19.980	9.990	4.995
4999	199.960	99.980	49.990	24.995
9999	399.960	199.980	99.990	49.995

sent with four different sending rates. The measured values (Table 4) can be compared to the ideal values (Table 5), describing how long the sending of the specified number of messages should take in an ideal system. The results show that the higher the sending rate and the more messages are to be sent, the longer the time needed.

Table 6 shows how much longer in percentage it took the middleware to send the fixed amount of messages, compared to the time it theoretically takes to send out the messages in an ideal system. The negative values in the first row of −0.7 and −0.1 % are caused by a possible error of 1 % in these measurements. The error is triggered by the implementation of the tests, more specific by the loop controlling how many messages have been sent. In this loop a timestamp is taken, just before a message is sent but not after. Thus the timestamp recorded between sending and receiving may vary by the duration of sending one message compared to the ideal system, which is in all cases 1 %.

Table 6 Increase in time with respect to the ideal conditions (in percentage)

Number of messages	Rate (messages per second)			
	25	50	100	200
99	−0.7	−0.1	4.2	8.4
499	0.0	1.5	4.0	9.7
999	0.1	1.8	3.6	10.1
4999	0.3	4.6	10.5	10.5
9999	0.4	4.5	9.1	20.1

Error 1 %

4.3 *Durability Tests*

4.3.1 Low Payload Durability Tests

This experiment analyses the effects a prolonged runtime has on the middleware. Messages with a payload of 64 bytes were being sent for 1, 10, 30 and 60 min with four different sending rates. Table 7 shows the amount of messages sent during the single runs of the experiment. For comparison with the theoretical maximum, the number of messages sent in an ideal system with the respective sending rates is shown in Table 8.

The results observed in Table 9 show how even in a long time of operation the middleware performs with throughputs only in one case barely under 90 % of the theoretical maximum. Using this payload of 64 bytes there is also no packet loss recorded.

4.3.2 High Payload Durability Tests

To further test the capabilities and to increase the burden on the middleware, for this long-time experiment the size of the payload was increased to 6400 bytes. Just like

Table 7 Number of received messages in a specific period of time using the middleware

Duration (min)	Rate (messages per second)			
	25	50	100	200
1	1428	2943	5825	10,914
10	14,947	29,297	57,801	109,034
30	44,936	88,481	173,427	324,371
60	89,852	176,716	348,860	646,759

Table 8 Ideal number of received messages in a specific period of time

Duration (min)	Rate (messages per second)			
	25	50	100	200
1	1500	3000	6000	12,000
10	15,000	30,000	60,000	120,000
30	45,000	90,000	180,000	360,000
60	90,000	180,000	360,000	720,000

Table 9 Percentage of messages sent compared to the ideal number

Duration (min)	Rate (messages per second)			
	25	50	100	200
1	95.2	98.1	97.1	91.0
10	99.6	97.7	96.3	90.9
30	99.9	98.3	96.3	90.1
60	99.8	98.2	96.9	89.8

Table 10 Number of received messages in a specific period of time using the middleware

Duration (min)	Rate (messages per second)			
	25	50	100	200
1	1481	2899	5513	10,197
10	14,806	28,981	55,228	101,572
30	44,385	86,948	165,690	306,146
60	88,803	173,828	331,075	613,034

Table 11 Percentage of messages sent compared to the ideal number

Duration (min)	Rate (messages per second)			
	25	50	100	200
1	98.7	96.6	91.9	85.0
10	98.7	96.6	92.0	84.6
30	98.6	96.6	92.1	85.0
60	98.7	96.6	92.0	85.1

before, messages were sent for 1, 10, 30 and 60 min, with four different sending rates. As a comparison, the number of messages sent in an ideal system with the respective rates can be used. The results observed in Table 10 show the number of messages sent during the experiment, while a look back to Table 8 shows the theoretical maximum.

As expected, the results (see Table 11) are worse when compared to sending a payload of only 64 bytes (see Table 9). But in the required sending rates of 100 messages per second and below, the performance still is greater than 90 % of the theoretical maximum.

5 Conclusions

In this work, a complete communication software solution, with fast tuning and demonstrated high performance, has been presented for Cooperating Object systems. It is especially appropriate in distributed system with real-time requirements.

Also, novel modules have been explained allowing the interconnection with other middlewares like ROS and the extension of the framework with any information type without difficulty.

It has then been shown how the developed ANIMO framework has been applied to two different projects with very different real-time communication needs showing the countless possibilities that the framework offers if it is applied to smart cities applications.

Finally, the results of a battery of tests have been presented to give the reader proof of the performance of the framework.

References

1. García-Valls, M., Basanta-Val, P., Estévez-Ayres, I.: Adaptive Real-Time Video Transmission Over DDS. Department of Telematics Engineering University Carlos III de Madrid, Leganés, Spain
2. Core Libraries and Utilities User's Manual of Real-Time Innovations, Version 4.5, http://www.rti.com
3. Xiong, M., Parsons, J., Edmondson, J., Nguyen, H., Schmidt, D.C.: Evaluating the Performance of Publish/Subscribe Platforms for Information Management in Distributed Real-time and Embedded Systems. Vanderbilt University, Nashville TN, USA
4. Official website of the ROS middleware, http://wiki.ros.org
5. The Apache Velocity Project, http://velocity.apache.org
6. Official website of the EC-SAFEMOBIL project, http://www.ec-safemobil-project.eu
7. Official website of VT MÄK, http://www.mak.com/
8. Official website of the ARCAS project, http://www.arcas-project.eu

SERAPH: Service Allocation Algorithm for the Execution of Multiple Applications in Heterogeneous Shared Sensor and Actuator Networks

Claudio M. de Farias, Wei Li, Flávia C. Delicato, Luci Pirmez, Paulo F. Pires and Albert Y. Zomaya

Abstract Shared Sensor and Actuator Networks (SSAN) represent a new design trend in the field of Wireless Sensor Networks (WSNs) that allows the sensing and communication infrastructure to be shared among multiple applications submitted by different users, instead of the original application-specific WSN design. In this paper, with the goal of fully utilising the network infrastructure and inspired by a service-oriented architecture, we modeled applications as sets of primitive services to be provided by sensor nodes. By using such approach, sensor nodes can perform different roles according to the services they offer and it is possible to identify common services required by different applications so that leveraging service sharing and optimizing the use of the network resources. With these premises, we propose an adaptive service selection and allocation algorithm called SERAPH that can efficiently utilise the underlying heterogeneous hardware resources, and yet

C.M. de Farias (✉) · F.C. Delicato · L. Pirmez · P.F. Pires
PPGI-iNCE, DCC-IM, Universidade Federal do Rio de Janeiro, Rio de Janeiro, Brazil
e-mail: claudiofarias@nce.ufrj.br

F.C. Delicato
e-mail: fdelicato@gmail.com

L. Pirmez
e-mail: luci@nce.ufrj.br

P.F. Pires
e-mail: paulo.f.pires@gmail.com

W. Li · A.Y. Zomaya
Centre for Distributed and High Performance Computing, School of Information Technologies, The University of Sydney, Sydney, NSW 2006, Australia
e-mail: liwei@it.usyd.edu.au

A.Y. Zomaya
e-mail: albert.zomaya@sydney.edu.au

A. Guerrieri et al. (eds.), *Management of Cyber Physical Objects in the Future Internet of Things*, Internet of Things,
DOI 10.1007/978-3-319-26869-9_5

provide the desired QoS level for multiple applications. Experimental results show that SERAPH provides competitive performance regarding energy efficiency, making it a promising task allocation algorithm for SSANs.

Keywords Wireless sensor networks · Shared sensor networks · Task allocation

1 Introduction

The Internet of Things (IoT) [1–3] has rapidly evolved in the last years as an umbrella term envisioning a world in which every single object (the so-called things) can be identified, controlled, and monitored through the Internet. These heterogeneous objects are endowed with sensing and/or actuation capabilities, and become able to capture data about physical variables, eventually process such data in order to provide valuable information, and also to act upon the physical world as a response to various stimuli. Smart objects communicate to each other and seamlessly collaborate with other physical and/or virtual resources available in the Internet to provide value-added information and new generation services for end-users.

Wireless sensor networks (WSN) [4, 5] are a key component of IoT infrastructures. Such distributed networks are composed of intelligent sensor nodes (a sub-set of smart objects), which work in a collaborative way, and are able to introduce or improve a wide variety of services to be available to humans and applications. However, in order to proper work as a core component of IoT, Wireless Sensor Networks need to evolve from their traditional application-specific design into a shared system design, where multiple applications simultaneously run on top of the deployed nodes. Such evolution in the WSN design has raised the emergent concept of Shared Sensor and Actuator Networks (SSANs) [6].

A recent research challenge of SSANs is how to allocate limited node resources to potential contending applications running on the same infrastructure. This issue has received notable attention from both academic and industry sides. One possible solution for tackling this challenge is to employ task allocation algorithms, which are responsible for looking for the best sensors (depending on the given criteria, such as processing power or residual energy) to perform tasks but also considering energy conservation of the network. By tasks, we mean the non-dividable units of execution that, when composed together, consist of an application submitted by users. Although several efforts have been made in the WSN field to develop task allocation algorithms, most of them focus on the case of running one or more applications in a single WSN with the main goal to achieve energy efficiency. Differently from WSN, task scheduling algorithms [7] designed for SSANs should not only pay close attention to the energy saving by tuning on/off different functional modules of sensor nodes in their lifetime, but should also fully exploit the common tasks from different applications so as to further improve the use of limited node resources. This is potentially efficient since tasks from multiple applications could run only once and the collected result be shared by all of them. Existing common tasks, if not properly addressed by the scheduling algorithm, will consume system energy in a less efficient way by repeatedly performing them. For instance, let us consider two applications running simultaneously on SSAN, a meeting detection application and a fire detection application [6]. A meeting detection application requires a presence sensor and a temperature sensor to analyze if the attendees are comfortable during a scheduled meeting. A fire detection application and a meeting application can share a temperature sensing task and a temperature threshold evaluation task.

Another concern that a task allocation algorithm must have is to assure the application's quality of service (QoS). A task allocation algorithm should be aware of each application QoS requirements and network conditions to select the best nodes that can satisfy the needs of application as much as possible.

The task scheduling approach [6–8] is one of the few early attempts to address the aforementioned challenges. It models the applications as task graphs and allocates the composite tasks to the candidate nodes that offer the requested functionalities in the system. Our proposal builds on the existing task scheduling solutions but goes a step further in the context of task representation. We argue that modeling an application as a service composition [9–13] is a well-suited approach, which abstracts the underlying heterogeneous sensor networks as a set of services that represent all the available functionalities [14] in a standardized and common way. Consequently, sensor nodes act as data providers and the set of WSNs working as a whole in a shared infrastructure acts as service provider for end users. Due to the well-known issues of limited hardware capabilities and energy resources in WSNs, each sensor node typically provides one or few non-divisible functionalities in its monitored area. In this paper, we call each kind of functionality as a primitive service. Since it is often unlikely for a single type of node to perform an application, we thus aim at using a group of different nodes to collectively provide services for applications. Each application is composed of a suitable combination of primitive services, which are selected from all the available services across WSNs.

WSNs operate in a highly dynamic environment. Sensor nodes often vary their operation status for energy conservation; wireless links are unstable and subject to several types of interferences, among other factors. In a SSN the dynamics is even higher, since new applications can arrive and running applications can terminate. Therefore, an adaptive service allocation approach [15] is required to make the provided services to adapt to the dynamic underlying execution context without explicit intervention from end-users at runtime. Moreover, it has to maintain the quality of service (QoS) at an acceptable level to prevent high quality reduction during the service time. In other words, the service allocation method should comprise a dynamic and QoS-aware mechanism that guarantees the services provision at desired level most of the time.

Given these premises, we focus on developing a service selection algorithm called SERAPH (SERvice allocation algorithm for multiple APplications execution in Heterogeneous SSANs) that efficiently utilises the underlying resources in service-oriented SSANs, and yet can provide a satisfactory level of QoS [16] from heterogeneous sensor nodes within a dynamic environment. In summary, this paper makes the following contributions:

1. An adaptive service-based approach is proposed to dynamically allocate applications (composition of services) to heterogeneous sensor nodes across multiple WSNs; the proposed solution is adaptive in the sense that it considers residual energy of nodes as well as the current values of delay and data loss in the network to make decisions on service allocation,
2. The allocation scheme is able to handle QoS requirements; current addressed QoS parameters are delay and packet loss,
3. Different roles are dynamically assigned to sensor nodes and taken into consideration when selecting the services for different applications; such roles denote the type of services a node is able to provide.

The remainder of the paper is organized as follows: Sect. 2 describes the related work and Sect. 3 introduces the models used in this work. Our proposed approach is detailed in Sect. 4. Experimental results in Sect. 5 demonstrate the effectiveness of our approach. Finally, Sect. 6 draws some conclusions and illustrates future work.

2 Related Work

Efstratiou et al. [17] was the first to propose the SSAN approach as an extension to the traditional design of WSAN. The idea behind the approach is tried to decouple the system infrastructure from application ownerships. A framework is created to allow WSAN infrastructure to be shared among multiple applications, which are potentially owned by different users. By achieving this goal, WSAN infrastructure is viewed as an accessible resource, which can be dynamically re-purposed and re-programmed by different authorities, in order to support multiple applications. Supported by the results presented in [17], FRESNEL is a recently launched SSN project focused on building a large scale federated sensor network with different applications sharing the resources from the same underlying hardware infrastructure. SenShare [6], as a part of the FRESNEL project, also presented an approach of constructing overlay sensor networks that are not only responsible for providing the most suitable members to perform an application, but also isolating the network traffic of this application from the traffic generated by other applications or the supporting mechanisms that are used to maintain the network overlay. For achieving the goal of traffic isolation, SenShare intentionally applies a 6 bytes long application routing header to each application packet, but the entire network message is still formatted under the IEEE 802.15.4 standard. The application routing header contains four sources of information, namely, app_id, seq_no, origin and destination, where app_id is the unique identification for each application, seq_no works along with other fields of the header to prevent the packet duplication, and origin and destination represent the sender and the receiver addresses respectively.

There are several works in the SSANs field that address the task allocation problem. In [18], the authors presented Utility-based Multi-application Allocation and Deployment Environment (UMADE), which is a task allocation system for distributing various applications based on QoM (quality of monitoring). QoM is a distributed quality metric for monitoring a physical phenomenon of interest that is based on the measurement accuracy. Thus, the quality of QoM depends on the monitoring performed by all nodes allocated to an application. Unlike traditional approaches that usually allocate nodes in the networks to a single application according to metrics such as the amount of resources used, delay, processing and power consumption, UMADE dynamically allocates nodes to multiple applications according to the application's QoM. An inherent property of an application is that sensing data belonging to sensors that are allocated to a same application are naturally correlated. As a consequence, the contribution of a sensor node for an application QoM is dependent of other sensor nodes allocated for the same application.

In [19], the authors propose a greedy algorithm to schedule applications onto a specific SSAN. The algorithm performs the task allocation taking into account the QoM of the applications. The applications' QoM depends on the node to which the application was allocated. Therefore, it is important that the allocation algorithm seeks to optimize the allocation among multiple applications of a SSAN to maximize the QoM. The proposed work uses the property of QoM sub-modularity. The QoM sub-modularity is due to the fact that the readings from the sensors of different nodes are often correlated. For example, once the temperature readings from different nodes in the same room are correlated with each other, the assignment of a new node in the room to perform the monitoring temperature does not produce a considerable QoM improvement.

The work of [20] is an extension of [18] and presents a distributed game-theoretic approach to application allocation in shared sensor networks. The authors transform the optimal application allocation problem to a sub-modular game and then develop a decentralized algorithm that only employs localized interactions among neighbouring nodes. The authors prove that the network can converge to a pure strategy Nash equilibrium with an approximation bound of 1/2. The authors validated their results through simulations based on three real-world datasets (Intel dataset, DARPA dataset and BWSN dataset) to demonstrate that their algorithm is competitive against a state-of-the-art centralized algorithm in terms of QoM. However, all of these approaches for task allocation are solutions that do not address the latency issue which often occurs in practical WSN applications.

In [21] the authors introduce a real-time trust management module for an auction based scheduling system that is able to validate the reliable bid value and determine faulty nodes and malicious entities. The main objective of this task allocation scheme is to maximize network lifetime by sharing tasks and network resources among applications, while enhancing the overall application quality of service (e.g., application deadline). The authors also propose a heuristic two-phase winner determination protocol to deal with the combinatorial reverse auction problem. Task allocation for wireless sensor networks with multiple concurrent applications (such as target tracking and event detection) requires sharing applications' tasks (such as sensing and computation) and available network resources. In this paper, the authors model the distributed task allocation problem for multiple concurrent applications by using a reverse combinatorial auction, in which the bidders (sensor nodes) are supposed to bid cost values (according to their available resources) for accomplishing the subset of the applications' tasks. Trust management schemes consist of a powerful tool for the detection of unexpected node behaviors (such as faulty or malicious). It is critical for participants (i.e., bidders and auctioneer) to estimate each other's trustworthiness before initiating the task allocation procedure.

The main difference between SERAPH and [18–21] is that SERAPH shares all kinds of tasks not only sensing tasks, what leads to better energy conservation.

3 Application and System Modeling

In this section, we present the application and system models used in our proposal.

3.1 Application Model

In our proposal, an application is defined as a 4-tuple $\langle S, G, \Delta t, Q \rangle$, where S denotes a list of ordered tasks and each task represents a required service. G is the geographic area, determined by four geographic coordinates that define the boundary of an area of interest which the application is required to monitor or sense. Δt is used to specify how long the application should last, and Q represents the QoS requirements of an application regarding its successful completion. Q is a 3-tuple $\langle D, L, E \rangle$ representing the application requirements in terms of delay (maximum end-to-end delay tolerated by an application), packet loss (maximum percentage of data loss) and energy consumption (percentage of remaining energy that a node must have to perform a service for this application). Those parameters were extracted from [22–25].

In the context of our work, task and service have the same meaning. A service indicates the action that can be performed by a sensor node so as to be requested to accomplish the relevant task. A task $s_i \in S$ has the following properties: $\langle \mathit{service\_type}, \mathit{SsR}, \mathit{SdR}, \mathit{TP}, \mathit{EC} \rangle$. The *service_type* element denotes the type of the service a task requests. Each task is associated with one and only one service type. We consider four types of services/tasks in our model: decision, routing, actuating and sensing. *SsR* and *SdR* are two user-defined elements to characterize the service provided by a node. *SsR* denotes the data receiving rate, meaning the time interval between consecutive data readings the service requires to be suitably performed. It has different meanings for each type of service. In a sensing service (for instance, sensing values of temperature, light intensity, humidity, acceleration, etc.) it represents the frequency of data readings, while in a decision service it represents the frequency of receiving readings as an input for the decision procedure. For a routing service, it represents the frequency of messages to be forwarded by a node. Finally, for an actuating service, it represents the frequency of messages that an actuator should receive to perform an action. *SdR* denotes the data sending rate to indicate the frequency of data sending during the time of service execution. The *TP* (*Time to perform*) element denotes how long a task is to be in execution inside the WSN [26]. In our work, Δt is the sum of the *TPs* of all tasks of an application. *EC* represents the amount of energy consumed by each task. This amount is different given each node platform. So, in order to calculate the amount of energy of each platform the node uses the model presented in [27].

The WSN applications considered in this paper consist of one or more tasks; thus an application may require a composition of services. Specifically, each of these tasks is the finest-grained and non-divisible element to constitute the WSN application. In this paper, we assume that tasks are heterogeneous. In this sense, we treat each application as a chained list where a task is dependent of the completion of the previous one. We also assume that once a task is initiated it cannot be interrupted until its completion (non-preemptive task).

3.2 System Model

We assume that there are k WSNs in the system denoted by the set $W = (W_1, W_2, \ldots, W_k)$, deployed into the target (monitored) area *M* to ensure that any event can be detected by at least one sensor node. We call *W* an SSAN. Any given WSN is modeled as an undirected graph G = (V, E), where V = $(v_1, v_2, \ldots, v_n)$ represents the set of sensor and actuator nodes (in this sense an actuator is only a node with different capabilities) and E = $(e_1, e_2, \ldots, e_m)$ represents the set of all possible communication links among the sensor nodes in the same WSN. Each WSN W_i has a single sink node SN_i and a number of sensor nodes V. Sink SN_i is used as a gateway between the WSNs and external networks (such as the Internet). The Sink Node dispatches user requests to the selected sensor nodes, which are able to provide the corresponding services in a given WSN. Sink maintains a data repository that describes the capabilities (services), the residual energy and the location information of each sensor node (we are assuming that each node is capable of announcing its location either by a GPS or a trilateration algorithm) [28]. Upon the completion of services, the sink is in charge of communicating the results to the applications. In addition, each sink SNi is also responsible for the communication with others sinks to exchange data/intermediate result whenever this is required. All nodes are synchronized having as base their correspondent *SN*.

Any sensor node can overlap the monitoring region of a set of other sensor nodes. This indicates that some regions of the monitored area W are mutually covered by multiple sensors that can perform a certain task. For any given sensor node v_i in *V*, *i* denotes the index of the sensor node that belongs to the WSN. A sensor node v_i is capable of providing one or more Services depending on their capabilities to collect/sample different types of data, for instance, temperature, light, smoke, and movement. Sensors in a same WSN are heterogeneous regarding their provided Services. All sensor nodes are endowed with the same radio interface as well as communication and sensing ranges. Moreover, sensors can detect all events of interest occurred within their sensing range.

According to the various services embedded on the nodes, we assign different roles to sensor nodes, namely: (i) relay nodes will receive only routing service; (ii) sensor nodes: can receive sensing and routing task; (iii) decision nodes: can receive a decision (any data processing task, such as a data fusion or any decision), sensing and routing task and (iv) actuators: actuating, decision, sensing and routing task. Please keep in mind, when the incoming user requests vary, the roles of the sensor nodes will be changed accordingly. In addition, we assume that all the sensors in the WSN have a valid communication path to reach their sink. Communication interference between sensor nodes is not considered in this paper.

4 SERAPH

The algorithm presented in this section aims at performing allocation of multiple applications over heterogeneous SSANs, while reacting to dynamic network conditions. Our proposed algorithm is designed as a centralized approach, with the sink node acting as a central decision node receiving information from all provisioning services and deciding on the services to be used. This sink node also acts as a coordinator when the underlying WSNs belong to different administrators so that the desired services can be properly invoked at the run-time.

SERAPH encompasses four phases: initialization, node filtering, service matching and service allocation. The pseudo-code of SERAPH is shown in Fig. 1.

First, in the initialization phase, records are created and populated in the sink node with data required to operate the system. The data structure called Application List is populated with the set of the 4-tuple $\langle S, G, \Delta t, Q \rangle$, as defined in Sect. 3. Another data structure called service List contains all services (as presented in Sect. 3).

In the node filtering phase, we create a list of candidate WSNs, meaning WSNs that have nodes able to execute the desired application according to the availability function described in Eq. 1:

$$A(s,x,y,i,c) = \begin{cases} 1, & \text{if a sensor is available} \\ 0, & \text{if a sensor is unavailable} \end{cases} \tag{1}$$

where s is one of the services that an application requires, x and y are the geographical location for the node (to check if the node is in the application's area of interest G, as defined in Sect. 3), i is the index of the sensor and c represents the role of the sensor. If the node's role set is equal to the service type than a node is fit for the task. Else the node is not available.

```
Input: Service based applications, topology of WSNs, application
requirements
Output: The service-sensor assignment
// Node Filtering
Consider Q as a queue for all the arrival applications
When a new application Ai arrives, decomposes the application Ai into n
services, where S_Ai ={S_1,S_2, … ,S_n}.
//node role assignment
FOR each WSN Wi in W
   FOR each node Vi in V
      FOR each Application Ai
         IF Vi has actuation capabilities then role = actuation
         IF Vi has sensing capabilities and then role = sensing
         IF Vi is neighbour to a node that has a sensing service in
               the same application then role = decision
         ELSE role = routing
      END FOR
   END FOR
END FOR
For each service in S_Ai look in the group of sensors V, using equation (1) to form
list of candidate sensors C_s
//Service Matching
FOR each service Si in S DO
FOR a given service S_i∈S_Ai, verify if this service is deployed in the WSN Wi.
IF Si is in Wi
take Si out of S
END IF
Put all the services a list Q
END FOR
//Service Allocation – In the sink node
FOR each candidate sensors list C_si
     FOR each Vj ∈ C_si DO
//The algorithm depicted on Figure 2 gives the parameters that will //be
used in equation (3).
     Select the sensor with the best value to the result generated by equation
(2) and that the role is equal to the service type
     End FOR
End FOR
Return the optimal sensor v_i for n_i
```

Fig. 1 The pseudo code of SERAPH

In the service matching phase, when a new application arrives, the sink node will analyse the composition of the application. First, SERAPH puts all newly arrived applications into a queue according to their arrival time. Then, it will search in the newcomer applications for services that are common to the applications that are already currently running in the system. Our algorithm excludes the common services from the list of services required by the new applications. In order to avoid losing data due to the reuse of a service [29], we have used the idea of session persistence presented in [27] and used in [4, 26]. The rationale behind the solution is to maximize the intersection of execution time of the same service from different applications. For example, suppose that two sensor nodes p1 and p2 are able to provide the same service S from a monitored area. A service S requested from application A1 is assigned to the sensor p1. When a new application A2 arrives, it contains the same service request for the same area. At the same time, the service S of application A1 is still running on S1. Instead of initializing a new sensor p2 to perform the task, we allocate this task to sensor p1, which is already providing the same service to application A1. Before the service S of application A1 stops running, the collected data can be used for both application A1 and A2. In the meantime, only one measurement of energy is drawn from the system, and the longer intersection time for the applications, the more energy the system saves.

In the service allocation phase, we evaluate, in the sink node, each service in face of the incoming application. We set the parameters of Eq. 2. For a given service, we derive the following Eq. 2:

$$\mathrm{SV} = O \times \mathrm{TP} \tag{2}$$

where SV means the sensor value (the value that will be used in the algorithm to select the best node to perform a task), TP means the topological distance, denoting the number of hops between the node and the sink node. O is an objective function that relates QoS parameters as it can be seen in Eq. 3:

$$O = \alpha \times D + \beta \times L + \gamma \times E \tag{3}$$

where α, γ and β are coefficients for the normalized ([0, 1]) delay, data loss and energy consumption, respectively. These three coefficients have a relationship $\alpha + \gamma + \beta = 1$. L represents the loss in percentage (%), the parameter $E = \frac{E_{\text{residual}} - E_{S_i}}{E_{\text{residual}}}$ represents the residual energy of the WSN. E_{S_i} denotes the energy consumption value for a given service. We use normalization because in a heterogeneous SSAN, sensors may use different battery models. D represents the delay of the service completion that is the duration of a given service, divided by the duration of a whole application (the sum of the duration of all services in an application).

The three considered QoS parameters (D, L and E—presented in Sect. 3) are: delay, data loss and energy consumption. The algorithm presented in Fig. 2 is used to find the coefficients of Eq. 3. This equation model was extracted from [30] and it is used due to its simplicity and low computational cost.

If Loss >=0.6 and Energy <=0.4 then α =0.3; β = 0.6; γ = 0.1;
If Loss <=0.4 and Energy >=0.6 then α = 0.6; β = 0.3; γ = 0.1;
If Delay <=0.4 and Energy >=0.6 then α = 0.4; β = 0.3; γ = 0.3;
If Delay >=0.6 and Energy <=0.4 then α = 0.3; β = 0.3; γ = 0.4;

Fig. 2 Coefficient generator

Once the best node is found using the latest information about nodes in the candidate sensor list, the sink node collects information about the allocated Services of application A_i for updating the system about the network conditions until the next arrival of application.

5 Experiments

The goal of the experiments is to assess: (i) SERAPH's efficient use of sensor resources (Sect. 5.3); (ii) SERAPH's capacity of satisfying user's QoS requirements (packet loss and delay) (Sect. 5.4) and (iii) SERAPH's adaptation capability (Sect. 5.5).

The experiments were conducted in the SUN SPOT platform [31], a commercial sensor platform that is particularly suitable for rapid development and demonstration of WSNs applications. The SUN SPOT SDK environment includes Solarium, that contains a SPOT emulator that is useful for testing software created using SUN SPOT and/or creating scenarios with a large number of nodes. In all the following experiments, we used a mix of sensor nodes (real nodes and virtual nodes emulated by Solarium) to ensure we can fully study how the proposed algorithm performs in large scale WSNs. Every experiment was repeated 30 times and had a confidence interval of 95 %.

5.1 Experimental Settings

In our experiments, we assumed that there are 1000 sensor nodes randomly distributed in a 1000 × 1000 m^2 field, organized as ten (10) independent WSNs (100 nodes each) with a single sink node placed in the top right corner for each one. In all experiments we varied the application number using 2, 4, 6, 8, and 10 applications. The values of QoS requirement of packet loss and delay are randomly chosen in a [1, 10 %] interval and [10–50 ms] interval respectively. The values of the intervals were chosen based on the applications presented in [32, 33].

As presented in Sect. 3, each application is composed of a set of services. In our experiments, each application contains 2–60 services. In order to generate a random set of services for each application, we have used a random graphs generation procedure further explained in [34]. All experiments were performed with 2 common services among applications.

We used LQRP as routing protocol to exchange data in each WSN. Each simulation runs for 900 s. We adopted the energy consumption model described in [27]. Interference issues are dealt by SUN SPOT MAC protocols and simulator and are out of the scope of our work. As synchronization protocol we have used the one presented in [31]. In our experiments, the area of interest of all applications corresponded to the same 1000 × 1000 m^2 where the nodes are deployed. We have compared SERAPH with a Naive Approach, the FRP algorithm [35] and SACHSEN algorithm [26]. We have created a naive approach that is a scheme for task scheduling as a baseline. In such scheme, each task is allocated in a random way that only checks if the node is able to execute the selected task (in terms of geographical position, required sensing interfaces and amount of energy to conclude the task). If the node randomly selected is not able to perform the task, then another node will be randomly chosen [36]. We have used the default noise generator from Sun SPOT.

5.2 Metrics

In our experiments, the network lifetime and the allocation successful rate (ASR) were the metrics used to evaluate SERAPH's efficient use of sensor resources (Sect. 5.3). The delay and packet loss are the metrics used for the evaluation the SERAPH's capacity of satisfying user's QoS requirements (packet loss and delay) (Sect. 5.4). In this paper, we adopted the same definition of network lifetime used in [26], which is the time elapsed until the first node in the WSN is completely depleted of its energy. As the network lifetime metric does not fully reveal the performance of SERAPH, we used the metric ASR (Allocation Success Rate). ASR was introduced in [18] and it is obtained from other two metrics: (i) the number of executed services and (ii) the number of missed services. The number of executed services is used to measure the total number of services that have been successfully executed in the experiment. The number of missed services denotes the number of services that fail to be performed due to various issues occurred in the network, such as data loss, sensor malfunctioning and loss of connectivity. Using these two metrics, we derive ASR as the ratio between executed services and the total number of services (a value between 0 and 1).

Delay is defined as the ratio between elapsed time between submitting an application into the SSAN and the result coming back to the Sink Node and the sum of all services composing an application. Packet loss is defined as the ratio of packets received by the Sink Node and the number of packets sent by source nodes.

The network lifetime and the allocation successful rate (ASR) were the metrics used to evaluate SERAPH's adaptation capability (Sect. 5.5).

5.3 Experiments Evaluating System Efficiency

We have performed a set of experiments in order to evaluate SERAPH's efficient use of sensor resources by varying the number of applications simultaneously running in different WSNs.

The first goal of this set of experiments is to assess how long the WSNs last using SERAPH and the other approaches considered in this work, by varying the

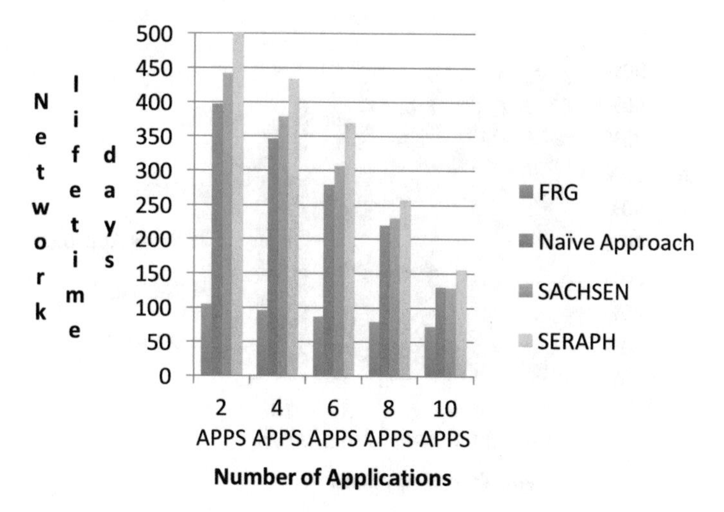

Fig. 3 Network lifetime with different number of applications

number of applications simultaneously running in different WSNs. The results of this experiment (Fig. 3) show that with the increment of the number of applications simultaneously running in WSNs, in all algorithms the network lifetime is reduced accordingly. FRG has the worst network lifetime performance among all algorithms. This is mostly due to the greedy design principle makes it to always search for all nodes that can perform an application request regardless of the overall system performance. SACHSEN and SERAPH showed better results in network lifetime performance than Naïve Approach and FRG since both use service sharing, which efficiently utilizes the hardware resources to execute WSN applications. The more common service requests are simultaneously being performed within the WSNs, the better performance the service-shared based algorithms will have. However, the network lifetime of SERAPH is better than SACHSEN's since SERAPH can adapt itself to the changing network conditions and, so, it has more chance to select a suitable node to perform the specific service from its role-assignment mechanism.

In the second set of experiments, we evaluate the ASR values for all algorithms by varying the number of applications simultaneously running in different WSNs. Figure 4 shows that as the number of applications increases, the ASR decreases for all algorithms. The Naive Approach showed the worst results among all algorithms. This occurred because the naive approach allocates tasks in a random way without considering the QoS requirements. FRG algorithm is superior to all other approaches, but this outcome is achieved sacrificing network lifetime to offer the best allocation. Especially, when the number of applications reaches 10, FRG gets a better performance than SERAPH (4 % better) but SERAPH has a much longer network lifetime (almost 2.3 times better).

SERAPH showed similar results to SACHSEN when the number of applications remains low in our experiments. However, our approach performed better when the number of applications starts increasing. SERAPH outperforms SACHSEN because

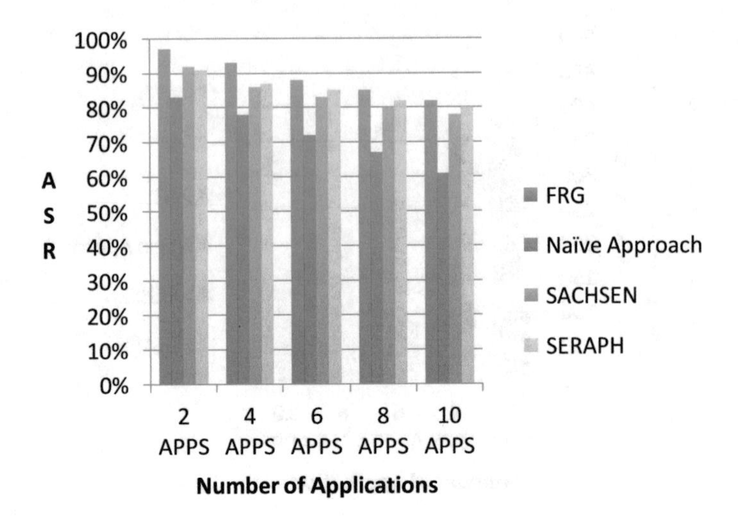

Fig. 4 ASR with a different number of applications

SERAPH considers QoS parameters in order to allocate the nodes. As the number of applications increases, also increases the number of messages sent, packets lost and delay. As the number of applications increases, it becomes more important for the allocation algorithm to consider the QoS parameters.

5.4 *Experiments Evaluating SERAPH Capacity to Meet QoS Requirements*

We have performed a set of experiments in order to evaluate SERAPH's capacity to satisfy QoS requirements of the application by varying the number of applications simultaneously running in different WSNs. In these experiments, we are trying to verify if SERAPH is capable to meet the applications requirements in terms of delay and packet loss.

First, we have performed an experiment to analyze if SERAPH is able to meet the packet loss requirement of each application as the number of applications is increased. We observed in Fig. 5 that as the number of applications simultaneously running in different WSNs increases the packet loss also increases.

FRG is the least affected by packet loss (it had the same result as SERAPH) since it is a greedy algorithm that chooses to allocate a task to a group instead of a single sensor. Therefore, in case of losing a sensor to perform a task, there is a replacement ready to be used. On the other hand, as seen in Fig. 3, this behavior leads to high-energy consumption.

Naive Approach and SACHSEN had the worst results since neither of them takes QoS into consideration in order to allocate nodes. The performance of

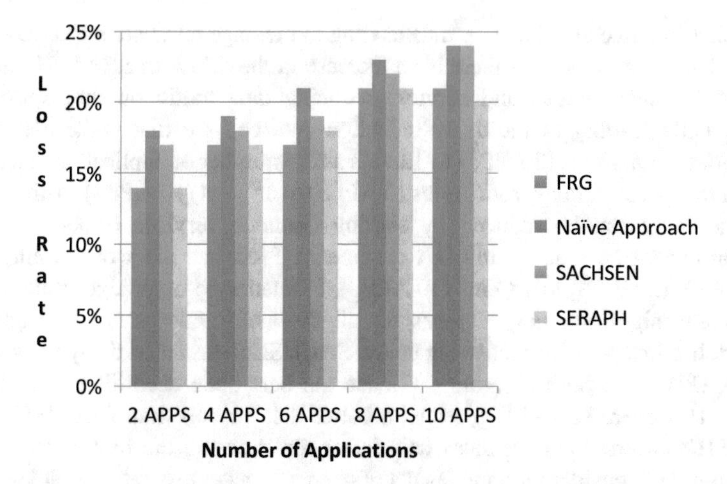

Fig. 5 Packet Loss with a different number of applications

SERAPH presents the best packet loss rate since it chooses the node to allocate the tasks based on the QoS requirements. This mechanism requires that up-to-date sensor node information is collected regularly. If any information of sensor nodes is lost on the way to the sink, those nodes are automatically out of the set of eligible nodes until such information becomes accessible again. Packet loss impact is therefore reduced to a certain extent, making the performance of SERAPH matches FRG (1 % of difference but with confidence interval of 2 %) with much higher System Lifetime (almost 2.3 times higher in worst cases).

The main goal of the second set of experiments is to verify if SERAPH is able to meet the delay requirements of each application as the number of applications increases. We can observe in Fig. 6 that SERAPH has a better performance in terms of delay compared to the other algorithms. As the number of applications simultaneously running in different WSNs increases, the data traffic and, consequently,

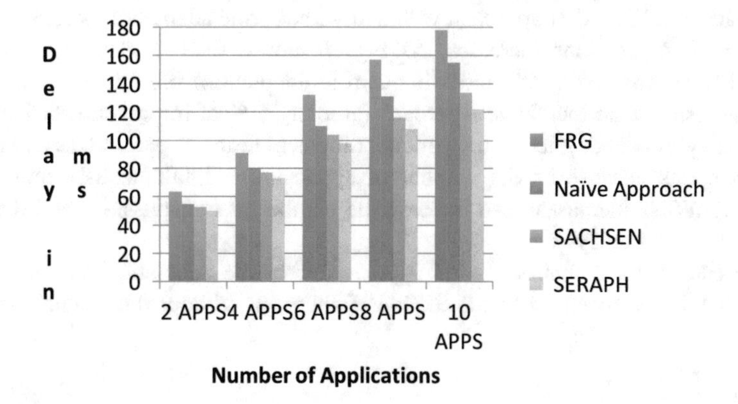

Fig. 6 Delay with a different number of applications

the packet loss rate also increase thus leading to message retransmissions or service failure. Due to these factors there is an increase in the delay. In SERAPH, multiple applications share nodes and services reducing data traffic on the WSNs and, consequently leading to the delay reduction. We can see that while SERAPH's delay goes from 48 ms (2 APPS) to 122 ms as the number of applications increases, the Naive Approach goes from 56 ms (2 APPS) to 157 ms (10 APPS). Although the network delay can be reduced by sharing common services (tasks), the tasks duration cannot be reduced in our experiments. FRG had the worst result, going from 64 ms (2 APPS) to 183 ms (10 APPS), in relation to other algorithms since it is a greedy algorithm that searches for all available nodes sacrificing not only network lifetime but also increasing delay. SACHSEN presented delay values close to SERAPH since both algorithms choose the best node according to their own criteria. However, SERAPH presents a better delay result than SACHSEN since SERAPH's criteria includes delay as a factor for choosing the best node.

Taking into consideration the results of average packet loss rate and delay, in the scenario with 10 applications, SERAPH was able to meet the specified QoS requirements of 92 % of applications while Naive Approach, FRG and SACHSEN were able to meet the specified QoS requirements of only 54, 81 and 84 % of applications respectively. This leads us to the conclusion that SERAPH can properly meet the applications requirements in terms of delay and packet loss.

5.5 Experiments Evaluating Adaptability

In order to evaluate the advantages of SERAPH's adaptability, we have performed a set of experiments by varying the number of applications simultaneously running in different WSNs. In this set of experiments, we evaluated SERAPH's efficient use of sensor resources (in terms of ASR and network lifetime) with and without the role adaptation feature. We have repeated the same set of experiment used in (Sect. 5.3).

As shown in Fig. 7, we observed that the network lifetime values for Naïve Approach and SERAPH approach with and without role adaptation decrease as the number of applications increases. As Fig. 7 shows, we can also see that both SERAPH approaches (with and without role adaptation) present better network lifetime results than the Naïve Approach (at least 25 % of improvement) thanks to its capacity of adapting to the network conditions in terms of packet loss and delay. This happens because as the number of applications simultaneously running in different WSNs increases, also increase the number of transmissions and thus the packet loss and delay.

We can also see that SERAPH approach with role adaptation presents better network lifetime results than the SERAPH approach without this feature (at least

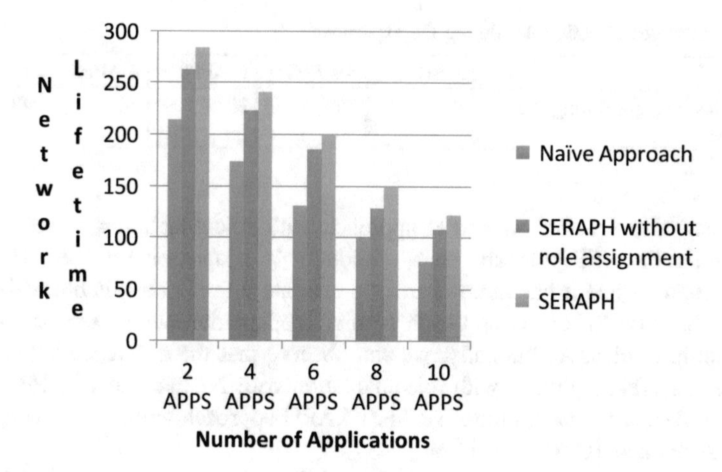

Fig. 7 Adaptability impact in terms of network lifetime

10 % of improvement) since the latter has more chance to select a suitable node to perform the specific service from its role-assignment mechanism.

As shown in Fig. 8, we observed that ASR values for Naïve Approach and SERAPH approaches with and without role adaptation decrease as the number of applications increases.

We can also see in Fig. 8 that both SERAPH approaches (with and without role adaptation) present better ASR than the Naïve Approach (at least 8 % of improvement) due to their capacity of adapting to the network conditions in terms of packet loss and delay. This happens because as the number of applications simultaneously running in different WSNs increases, the number of transmissions increases and thus the packet loss and delay is also increased.

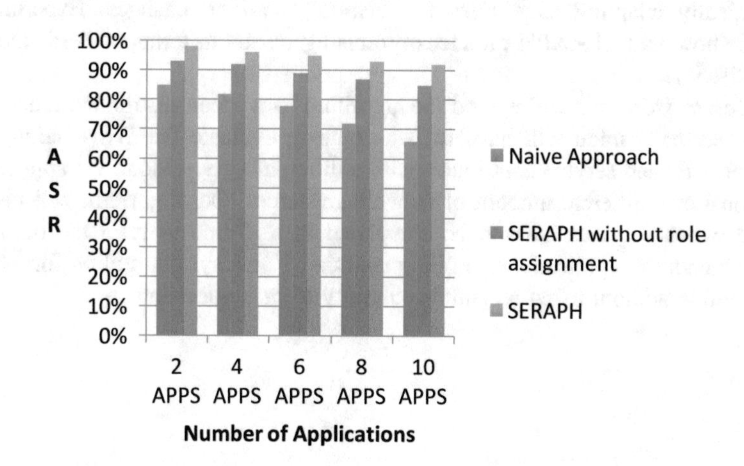

Fig. 8 Adaptability impact in terms of ASR

Table 1 Changes of node roles during the experiment

	2 APPS	4 APPS	6 APPS	8 APPS	10 APPS
% of roles changes during the experiment	8	15	23	38	48

We can also see that SERAPH approach with role adaptation presents better ASRs than the SERAPH approach without role adaptation (at least 10 % of improvement) since it has more chance to allocate a task to a suitable node. This happens because SERAPH approach with role adaptation allows sensor nodes to have multiple roles. Additionally, we can observe that the difference between the ASR in SERAPH approach with role adaptation with 2 APPs and 10 APPs is 4 % and the difference between the ASR in SERAPH approach without role adaptation with 2 APPs and 10 APPs is 11 %.

Another interesting conclusion in Table 1 is how the role adaptation is important to allocate tasks successfully. In the experiments, if we consider the percentage of nodes that have changed their roles to accommodate the arriving applications, we are able to see how important the adaptability is to SERAPH. As the application number increases, the percentage of changes in nodes roles is increased from 8 %, with 2 APPS, to 48 % with 10 APPS.

6 Conclusion

In this paper, we presented a service selection and allocation algorithm called SERAPH for execution of multiple applications in heterogeneous service-oriented SSNs. SERAPH exploits common services to make energy-efficient service-sensor assignments. In addition, it explicitly takes full advantage of different sensor roles, dynamically adapting such roles to network condition changes. Experimental results show that SERAPH produces promising results in terms of ASR and network lifetime.

In future work, we will extend the algorithm to work in distributed manner. In such case, each node will have to contribute to evaluate the WSN, taking into account that each service contributes with a different QoS value to the application, and consumes different amount of network resources. Our algorithm will need to decide which services have to be completed in real-time to meet the minimum requirements of applications, while maximizing the system utility for all the applications without using a centralized entity to make decisions.

References

1. Gubbi, J., Buyya, R., Marusic, S., Palaniswami, M.: Internet of things (IoT): a vision, architectural elements, and future directions. Future Gener. Comput. Syst. **29**(7), 1645–1660 (2013)
2. Billet, B., Issarny, V.: From task graphs to concrete actions: a new task mapping algorithm for the future internet of things. In: 11th IEEE International Conference on Mobile Ad Hoc and Sensor Systems (MASS), pp. 470–478, Oct 2014
3. Fortino, G., Guerrieri, A., Russo, W., Savaglio, C.: Middlewares for smart objects and smart environments: overview and comparison. In: Internet of Things Based on Smart Objects, pp. 1–27. Springer, Berlin (2014)
4. Li, W., Delicato, F.C., Zomaya, A.Y.: Adaptive energy-efficient scheduling for hierarchical wireless sensor networks. ACM Trans. Sen. Netw. **9**(3), 1–34 (2013)
5. Fortino, G., Giannantonio, R., Gravina, R., Kuryloski, P., Jafari, R.: Enabling effective programming and flexible management of efficient body sensor network applications. Hum. Mach. Syst. IEEE Trans. **43**(1), 115–133 (2013)
6. Leontiadis, I., Efstratiou, C., Mascolo, C., Crowcroft, J.: SenShare: transforming sensor networks into multi-application sensing infrastructures. In: Picco, G., Heinzelman, W. (eds.) Wireless Sensor Networks, pp. 65–81. Springer, Berlin (2012)
7. de Farias, C.M., Pirmez, L., Delicato, F.C., Li, W., Zomaya, A.Y., De Souza, J.N.: A scheduling algorithm for shared sensor and actuator networks. In: IEEE International Conference on Information Networking (ICOIN), pp. 648–653, Jan 2013
8. Li, W., Delicato, F.C., Pires, P.F., Zomaya, A.Y.: Energy-efficient three-phase service scheduling heuristic for supporting distributed applications in cyber-physical systems. In: Proceedings of the 15th ACM International Conference on Modeling, Analysis and Simulation of Wireless and Mobile Systems, Paphos, Cyprus, pp. 229–238 (2012)
9. Geyik, S., Szymanski, B., Zerfos, P.: Robust dynamic service composition in sensor networks. Serv. Comput. IEEE Trans. 6(4), 1–1 (2012)
10. Bell, M.: SOA Modeling Patterns for Service Oriented Discovery and Analysis. Wiley, London (2010)
11. Cheng, B.C., Lemos, R., Giese, H., Inverardi, P., Magee, J., Andersson, J., Becker, B., Bencomo, N., Brun, Y., Cukic, B., Marzo Serugendo, G., Dustdar, S., Finkelstein, A., Gacek, C., Geihs, K., Grassi, V., Karsai, G., Kienle, H., Kramer, J., Litoiu, M., Malek, S., Mirandola, R., Müller, H., Park, S., Shaw, M., Tichy, M., Tivoli, M., Weyns, D., Whittle, J.: Software engineering for self-adaptive systems: a research roadmap. In: Cheng, B.C., Lemos, R., Giese, H., Inverardi, P., Magee, J. (eds.) Software Engineering for Self-Adaptive Systems, pp. 1–26. Springer, Berlin (2009)
12. Rouvoy, R., Eliassen, F., Floch, J., Hallsteinsen, S., Stav, E.: Composing components and services using a planning-based adaptation middleware. In: Pautasso, C., Tanter, É. (eds.) Software Composition, pp. 52–67. Springer, Berlin (2008)
13. Fortino, G., Guerrieri, A., Russo, W.: Agent-oriented smart objects development. In: IEEE 16th International Conference on Computer Supported Cooperative Work in Design (CSCWD), pp. 907–912, May 2012
14. Rao, J., Su, X.: A survey of automated web service composition methods. In: Cardoso, J., Sheth, A. (eds.) Semantic Web Services and Web Process Composition, pp. 43–54. Springer, Berlin (2005)
15. Fok, C.-L., Roman, G.-C., Lu, C.: Adaptive service provisioning for enhanced energy efficiency and flexibility in wireless sensor networks. Sci. Comput. Program. **78**(2), 195–217 (2013)

16. Paschoalino, R., Madeira, E.R.M.: A scalable link quality routing protocol for multi-radio wireless mesh networks. In: Proceedings of 16th International Conference on Computer Communications and Networks (ICCCN), pp. 1053–1058, 13–16 Aug 2007
17. Efstratiou, C., Leontiadis, I., Mascolo, C., Crowcroft, J.: A shared sensor network infrastructure. In: Proceedings of the 8th ACM Conference on Embedded Networked Sensor Systems (SenSys'10), pp. 367–368. ACM, New York (2010). doi:10.1145/1869983.1870026
18. Xu, Y., Saifullah, A., Chen, Y., Lu, C., Bhattacharya, S.: Near optimal multi-application allocation in shared sensor networks. In: Proceedings of the Eleventh ACM International Symposium on Mobile Ad Hoc Networking and Computing, Chicago, pp. 181–190 (2010)
19. Bhattacharya, S., Saifullah, A., Lu, C., Roman, G.-C.: Multi-application deployment in shared sensor networks based on quality of monitoring. In: 2010 16th IEEE Real-Time and Embedded Technology and Applications Symposium, pp. 259–268 (2010). doi:10.1109/RTAS.2010.20
20. Wu, C., Xu, Y., Chen, Y., Lu, C.: Submodular game for distributed application allocation in shared sensor networks. In: Proceedings of IEEE Conference on INFOCOM, pp. 127–135, Mar 2012
21. Edalat, N., Xiao, W., Motani, M., Roy, N., Das, S.K.: Auction-based task allocation with trust management for shared sensor networks. Secur. Commun. Netw. **5**(11), 1223–1234 (2012)
22. Siu-Nam, C., Chan, A.T.S.: Dynamic QoS adaptation for mobile middleware. Softw. Eng. IEEE Trans. **34**(6), 738–752 (2008)
23. Floch, J., Hallsteinsen, S., Stav, E., Eliassen, F., Lund, K., Gjorven, E.: Using architecture models for runtime adaptability. Softw. IEEE **23**(2), 62–70 (2006)
24. Geihs, K., Barone, P., Eliassen, F., Floch, J., Fricke, R., Gjorven, E., Hallsteinsen, S., Horn, G., Khan, M.U., Mamelli, A., Papadopoulos, G.A., Paspallis, N., Reichle, R., Stav, E.: A comprehensive solution for application-level adaptation. Softw. Pract. Experience **39**(4), 385–422 (2009)
25. Fortino, G., Guerrieri, A., O'Hare, G.M., Ruzzelli, A.: A flexible building management framework based on wireless sensor and actuator networks. J. Netw. Comput. Appl. **35**(6), 1934–1952 (2012)
26. Li, W., Delicato, F.C., Pires, P.F., Lee, Y.C., Zomaya, A.Y., Miceli, C., Pirmez, L.: Efficient allocation of resources in multiple heterogeneous wireless sensor networks. J. Parallel Distrib. Comput. **74**(1), 1775–1788 (2014)
27. Li, W., Vesilo, R.: Modeling of session persistence in web server systems. In: Australian Telecommunications Networks and Application Conference, Melbourne (2006)
28. Delicato, F., Protti, F., Pirmez, L., de Rezende, J.F.: An efficient heuristic for selecting active nodes in wireless sensor networks. Comput. Netw. **50**(18), 3701–3720 (2006)
29. Garlan, D., Shang-Wen, C., An-Cheng, H., Schmerl, B., Steenkiste, P.: Rainbow: architecture-based self-adaptation with reusable infrastructure. Computer **37**(10), 46–54 (2004)
30. Herrera, F., Herrera-Viedma, E.: Linguistic decision analysis: steps for solving decision problems under linguistic information. Fuzzy Sets Syst. **115**(1), 67–82 (2000)
31. Reyes, J.A.G., Robles, R.S., Recéndez, B.E.S., Olague, J.G.A.: Implementation of a timestamping service for SunSPOT sensors. Procedia Technol. **7**, 4–10 (2013). ISSN 2212-0173, http://dx.doi.org/10.1016/j.protcy.2013.04.001
32. Rowaihy, H., Johnson, M.P., Liu, O., Bar-Noy, A., Brown, T., Porta, T.L.: Sensor-mission assignment in wireless sensor networks. ACM Trans. Sen. Netw. **6**(4), 1–33 (2010)
33. Noh, A.S.-I., Lee, W.J., Ye, J.Y.: Comparison of the mechanisms of the zigbee's indoor localization algorithm. In: International Conference on Software Engineering, Artificial Intelligence, Networking, and Parallel/Distributed Computing, 2008 (SNPD'08), pp. 13, 18, 6–8 Aug 2008. Ninth ACIS

34. Xiong, S., Li, J., Li, M., Wang, J., Liu, Y.: Multiple task scheduling for low-duty-cycled wireless sensor networks. In: INFOCOM'11
35. Ting, Z., Mohaisen, A., Yi, P., Towsley, D.: DEOS: dynamic energy-oriented scheduling for sustainable wireless sensor networks. In: Proceedings of IEEE on INFOCOM, pp. 2363–2371, 25–30 Mar 2012
36. Shuguang, X., Jianzhong, L., Zhenjiang, L., Jiliang, W., Yunhao, L.: Multiple task scheduling for low-duty-cycled wireless sensor networks. In: Proceedings of IEEE on INFOCOM, pp. 1323–1331, 10–15 April 2011

A Smart Platform for Large-Scale Cyber-Physical Systems

Andrea Giordano, Giandomenico Spezzano and Andrea Vinci

Abstract Recent advancements in the fields of embedded systems, communication technologies and computer science, have laid the foundations for new kinds of applications in which a plethora of physical devices are interconnected and immersed in an environment together with human beings. These so-called Cyber-Physical Systems (CPS) issue a design challenge for new architecture in order to cope with problems such as the heterogeneity of devices, the intrinsically distributed nature of these systems, the lack of reliability in communications, etc. In this paper we introduce Rainbow, an architecture designed to address CPS issues. Rainbow hides heterogeneity by providing a Virtual Object (VO) concept, and addresses the distributed nature of CPS introducing a distributed multi-agent system on top of the physical part. Rainbow aims to get the computation close to the sources of information (i.e., the physical devices) and addresses the dynamic adaptivity requirements of CPS by using Swarm Intelligence algorithms.

1 Introduction

The increasing use of smart devices and appliances opens up new ways to build applications that integrate the physical and virtual world into consumer-oriented context-sensitive Cyber-Physical Systems (CPS) [1–3] enabling novel forms of interaction between people and computers. CPS are combinations of physical entities controlled by software systems to accomplish specified tasks under stringent real-time and physical constraints.

A. Giordano (✉) · G. Spezzano · A. Vinci
CNR-ICAR, Via P. Bucci 41C, 87036 Rende, CS, Italy
e-mail: giordano@icar.cnr.it

G. Spezzano
e-mail: spezzano@icar.cnr.it

A. Vinci
e-mail: vinci@icar.cnr.it

A. Guerrieri et al. (eds.), *Management of Cyber Physical Objects in the Future Internet of Things*, Internet of Things,
DOI 10.1007/978-3-319-26869-9_6

The emerging cyber-physical world interconnects a vast variety of static and mobile resources, including computing/medical/engineering devices, sensor/ actuator networks, swarm of robots etc. Examples of CPS applications include [4] traffic control, power grid, smart structures, environmental control, critical infrastructure control, water resources and so on. These systems could be pervasively instrumented with sensors, actuators and computational elements to monitor and control the whole system. Furthermore, these devices should be interconnected so as to communicate and interact with each others and with people.

This scenario is supported by recent technology advancement in the fields of communication, embedded systems and computer science.

The networked cyber-physical world has a great potential for achieving tasks that are far beyond the capabilities of existing systems. However, the problem of effectively composing the services provided by cyber and physical entities to achieve specific goals remains a challenge [2, 3, 5]. Advanced models and architectures, autonomous resource management mechanisms, and intelligent techniques are needed for just-in-time assembly of resources into desired capabilities.

The complexity of a CPS, and the large number of elements involved, makes data analysis and operation planning a very difficult task. A currently used approach involves two layers: a *physical* layer and a *remote* (cloud) cyber layer. The physical layer sends sensed data to a remote server, which processes them and computes a suitable operation plan. Afterwards, the remote server sends the sequence of operations it must execute to each device on the physical layer. The reasoning is performed in the remote layer. This solution cannot be applied when there are constraints on *responsivity* time, that is, when a system needs to react fast to critical events that may overwhelm its integrity and functionality. Communication lag and remote processing can cause delays that a system simply cannot bear.

A wide variety of applications means a wide variety of devices. Currently, there is a plethora of different devices, each with its own particular functionalities and capabilities. There are simple devices without any computational unit as well as "smarter" devices with high computation power inside. There are devices with no operating systems and devices with simple or complex operating systems, such as tinyOS or Android. Our framework is designed to cope with this inherent heterogeneity.

To address the issues described above, our proposal moves on these main lines:

- Hiding the heterogeneity of CPS by introducing a *virtual object* layer.
- Moving the computation as close as possible to the physical resources in order to foster good performance and scalability.
- Introducing a distributed intelligence layer between the physical world and remote servers (cloud), which can execute complex tasks and horizontally/ vertically coordinate the devices;
- Switching from a cloud-based model to a cloud-assisted one, where the intelligent intermediate level carries out almost all the real-time control tasks, whereas the remote cloud level remains in charge of non-real-time tasks such as offline data analysis or presentation. The information provided by the data

analysis executed by the remote server are used by the intermediate level to optimize its operations and behaviour.

In this paper we propose a three-tier architecture (Rainbow) that uses single-board computers such as the Raspberry PI to connect massive-scale networks of sensors. This architecture is composed by the *Cloud* layer, the *Intermediate* layer and the *Physical* layer. Sensors are partitioned into groups, each of which is managed by a single computing node. These computing nodes host multi-agent applications designed to monitor multiple conditions or activities within a specific environment.

We present a new integrated vision that allows the designing of a large-scale networked CPS based on the decentralization of control functions and the assistance of cloud services to optimize their behaviour. Decentralization will be obtained using a distributed multi-agent system in which the execution of a CPS application is carried out through agents' cooperation [6–9]. The distributed multi-agent system lays the foundations for properly exploiting swarm intelligence concepts. Swarm intelligence [10, 11] systems are typically multi-agent systems made up of a population of simple agents interacting locally with one another and with their environment. The agents follow very simple rules, and although there is no central control structure dictating how individual agents should behave, local and to a certain degree random, interactions among such agents lead to the emergence of "intelligent" global behaviour, unknown to the individual agents. Natural examples of swarm intelligence include ant colonies, bird flocking, animal herding, bacterial growth, and digital infochemicals. Agents interacting with cloud services can exploit the analysis, predicting, optimization and mining scalable capabilities on historical data allowing applications to adjust their behaviour to best optimize their performance.

The remainder of this paper is structured as follows: Sect. 2 summarizes the current literature about cyber physical systems and the approaches to cope with its issues; Sect. 3 is devoted to a description of the proposed Rainbow Architecture; Sect. 4 describes two example of use; finally, we draw conclusions and the future works.

2 Related Work

In the recent years, the world has witnessed a real revolution about people habit in terms of ability of exploiting high technology solution in everyday life. This new scenario opens up new challenges regarding how the physical stuff can be used and integrated with the preexisting digital world. In these so called Cyber-Physical systems, many physical components collaborate each other by means of network communications in order to sense and act upon the physical world. These physical components are enhanced by using computational resources which supplies them the "smartness" needed to cope with complex tasks as controlling the physical environment and supporting most of the everyday human activities [12]. This scenario is supported by recent technology advancement in the fields of communication technologies, embedded systems and computer science. On the communication technologies side, new protocols like EPC TDS and IPv6 ensure unique addressability for all the elements involved in a CPS, while connectivity technologies like IEEE 802.11, ZigBee, Umts and ZTE, ensure light and fast connection both among the devices and between the devices and the Internet. On the embedded systems side, the miniaturization and the constant improvement of energy efficiency of electronic components enables the environment to be easily instrumented with sensors, actuators and computing devices, while the presence on the market of cheap and general purpose single-board computers, like Raspberry PI [13] and BeagleBoard opens up to new approaches and application scenarios. Finally, on the computer science side, the development of new techniques to analyse a massive volume of data, together with the advances in the fields of artificial and swarm intelligence, allows us to properly deal with even a large number of devices.

The inherent complexity of these kinds of systems is highlighted in [4], where cyber-physical issues are summarized in the following topics

- integrate the physical components in the digital world;
- supply each physical component of its own computational capabilities;
- communication and networking issues;
- dynamic reconfiguration;
- human interactions;
- security and reliability.

In the current literature these kinds of issues are addressed by two parallel research communities, the first concerning the Cyber-Physical Systems and the second related to the Internet of Things (IoT) vision [1]. Both research fields focus on the integration between physical and digital world but with different approaches and visions. The first research community, mainly from USA, comes from the control theory and control systems engineering fields and focuses on how the physical components can be interconnected each other and exploited using complex software entities. The second, the IoT community, is mainly driven from computer science field and Internet technologies and focuses mainly on heterogeneity and interoperability issues which comes from the integration of the physical components in the pre-existing Internet.

Formal distinction between CPS and IoT is yet to come [1], indeed, both share similar visions and are often used in the current literature to identify the same topics.

In order to deal with architectural issues posed by CPS, and well highlighted in [2], several frameworks and architectures have been proposed. Dilon et al. [14] proposes Web of Things based framework for CPS where physical sensors and actuators are universally identified (through URIs) and modelled as WoT resources. The resources can be accessed by means of remote RESTful invocations. The architecture is built using two key components, CPS Node and CPS Fabric. CPS nodes are directly connected with physical devices and interconnected each other using the CPS fabric, which consists of a set of network system functions (i.e. routing, admission control, and so on). An Intelligent Vehicle System is shown as a case of study.

In [15–17] some middlewares are proposed which implement the pervasive computing paradigm in the CPS context. Each one presents a different framework which allows programmers to design and develop applications by coding in an high level of abstraction. The code is automatically compiled and deployed to the suitable computing nodes where sensors and actuators are managed through a wireless network. In [16], the framework provides an object oriented model and a rule-based mechanism. The framework also introduces the concept of *item* which consists in physical entities dynamically detected by the system on the basis of the values measured by the sensors. Every time a physical item is identified by a specific set of rules, an associated software object is properly instanced. Hnat et al. [15] proposes a similar approach but exploiting Matlab.

Several works deal with CPS issues adopting multi-agent paradigms. In [18] a multi-agent system is used together with an event-based mechanism. In [9], a framework is described which lies on the *semantic agent* concept while in [8] the multi-agent systems is integrated with a service-oriented architecture (SOA). This work also suggests how well-known and proven swarm intelligence techniques can be properly adopted for industrial purposes. At last, in [7], an agent-based middleware for cooperating smart-objects is proposed. An implementation using JADE is also provided where a topic-based publish/subscribe protocol is exploited for permitting cooperation among agents.

3 Rainbow Architecture

Rainbow is a three-layer architecture designed in order to bring the computation (i.e. the controlling part) as close as possible to the physical part. Since CPS foresees that physical entities are spread across a large (even geographic) area, the previous assumption implies the controlling part to be intrinsically distributed.

Our proposal foresees the use of a *distributed agent-based* layer in order to address the aforementioned issues. The agent paradigm has several important characteristics:

Autonomy. Each agent is self-aware and has a self-behaviour. It perceives the environment, interacts with others and plans its execution autonomously.

Local views. No agent has a full global view of the whole environment but it behaves solely on the basis of local information.

Decentralization. There is no "master" agent controlling the others, but the system is made up of interacting "peer" agents.

Through these basic features, multi-agent systems make it possible to obtain complex *emergent* behaviours based on the interactions among agents that have a simple behaviour. Examples of emergent behaviour could refer to the properties of adaptivity, fault tolerance, self-reconfiguration, etc. In general, we could talk about *swarm-intelligence* when an "intelligent" behaviour emerges from interactions among simple entities. There is a plethora of bio-inspired swarm intelligence approaches in the literature that could be properly adopted in the context of CPS. In Sect. 4.2 we show an example where Swarm Intelligence is used to map noise pollution inside a city area.

Rainbow architecture is shown in Fig. 1. As it can be seen, the architecture could be divided into three layers. The bottom layer is the one that is devoted to the physical part. It encloses sensors and actuators, together with their relative computational capabilities, which are directly immersed in the physical environment.

In the Intermediate layer, sensors and actuators of the physical layer are represented as virtual objects (VOs). VOs offer to the agents a transparent and ubiquitous access to the physical part due to a well-established interface exposed as API. VO allows agents to connect directly to devices without care about proprietary drivers or addressing some kind of fine-grained technological issues. Each VO comprises "functionalities" directly provided by the physical part. Essentially, a VO exposes an abstract representation (i.e. *machine readable-description*) of the features and capabilities of physical objects spread in the environment. Functionalities exposed by different types of VOs can be combined in a more sophisticated way on the basis of event-driven rules which affect high-level applications and end-users.

In summary, as detailed in Sect. 3.1, all the devices are properly wrapped in VOs which, in turn, are enclosed in distributed *gateway* containers. The computational nodes that host the gateways represent the middle layer of the Rainbow architecture. Each node also contains an agent server that permits agents to be executed properly. Gateways and agent servers are co-located in the same computing nodes in order to guarantee that agents exploit directly the physical part through VO abstraction.

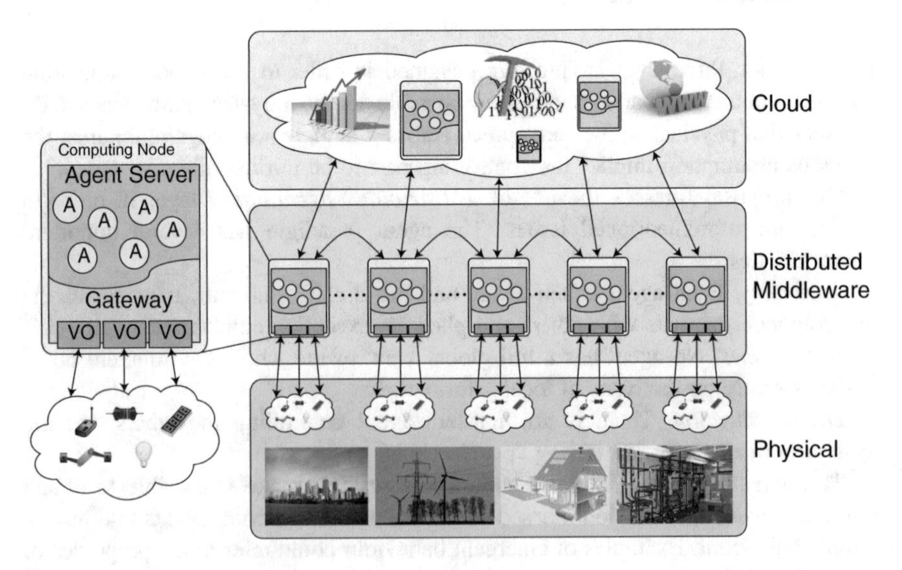

Fig. 1 Rainbow architecture

Instead of transferring data to a central processing unit, we actually transfer the process (i.e. fine-grain agent's execution) toward the data sources. As a consequence, less data needs to be transferred over a long distance (i.e. toward remote hosts) and local access and computation will be fostered in order to achieve good performance and scalability.

The upper layer of Rainbow architecture concerns the cloud part. This layer addresses all the activities that cannot be properly executed in the middle layer like, for instance, algorithms needing complete knowledge, tasks that require high computational resources or when a historical data storage is mandatory. On the contrary, all tasks where real time access to the physical part is required could be suitably executed in the middle layer.

3.1 Virtual Objects

We address issues about heterogeneity in CPS by introducing the Virtual Object (VO) concept. VO aims to hide heterogeneity by supplying a well-established interface permitting the physical parts to be suitably integrated with the rest of the system.

VO could be defined as a collection of physical entities like sensors and actuators, together with their computational abilities.

It can be composed of just a simple sensor or it can be a more complex object that includes many sensors, many actuators, computational resources like CPU or memory and so on.

In general, VO outputs can be represented by *punctual values* (e.g. the temperature at a given point of a room) or *aggregate values* (e.g. the average of moisture during the last 8 h). Also, the values returned by VOs could be just the measurement of sensors or could be the result of complex computations (e.g. the temperature of a given point of space computed by means of interpolation of the values given by sensors spread across the environment).

Furthermore, a VO could supply actuation functionality by changing the environment on the basis of external triggers or internal calculus.

These different kinds of behaviour that VO can expose must be taken into account. VO is therefore conceived as a complex object that can read and write upon many simple physical resources. More in detail, we consider that each VO exposes different *functionalities*. Each functionality can be either sensing or actuating and can be refined by further parameters that dynamically configure it.

The previous assumption leads to the definition of *resource* as the following *triplet*:

$$[VOId, VOFunctionId, Params]$$

where `VOId` uniquely identifies the VO, `VOFunctionId` identifies the specific functionality and `Params` is an ordered set of parameter values that configure the functionality.

For example let's consider a *Virtual Room* made of sensors for measuring different physical quantities inside a room such as *temperature*, *moisture*, *brightness* and so on. Suppose now you want to read from Smart Room the temperature in a given spatial point of the room. In that case the triplet could be:

$$[VirtualRoom, temperature, [x, y, z]]$$

where `x`, `y` and `z` are the cartesian coordinate of the point of interest.

Using object oriented terminology, a Resource could be seen as a particular "instance" of a functionality of a given VO.

Besides read and write operations (i.e. sensing and actuation), it is provided for VOs to be able to manage events that occur in the physical part. To that scope, our proposed middleware includes a *publish/subscribe* component for managing events in each computing node. Each event is defined by a *logical rule* where one or more VOs could be involved.

Each rule is a *logical proposition* in which the *atomic predicates* can be of the following kinds:

- *resource < threshold (e.g. temperature <300)*
- *resource > threshold*
- *boolean_resource (e.g. the_door_was_unlocked)*

Just an example of rule:

$$(temperature < 100 \textit{ and } brightness > 500) \textit{ or } people > 3 \textit{ or } door\_unlocked$$

The publish/subscribe manager component is in charge to parse the logical rule and generate a *binary tree* made as explained below: each node N of the tree corresponds to a logical proposition $N()$. Given L and R, the child nodes of N, their associated logical propositions are respectively $L()$ and $R()$ so that it results either $N() = L()$ *or* $R()$ or $N() = L()$ *and* $R()$. The radix of the tree corresponds to the entire rule while the leaves contain the atomic propositions that is passed to the suitable VOs. A binary tree representation example of a composed rule is shown in Fig. 2.

A VO is in charge to establish each time when the assigned atomic propositions are *true* or *false*. The logical proposition of a given node is computed on the basis of the value of its child nodes. The root of the tree is recursively involved by this bottom–up computation. As soon as the value of the root node (i.e. the value of the entire rule) changes all the subscriber will be notified.

All the physical things linked to a computing node together with relative VOs is enclosed in the *Gateway* container. The Gateway exposes an interface to interact directly with the VOs.

Each gateway represents the "entry point" that agents can use to exploit VOs of the relative computing node.

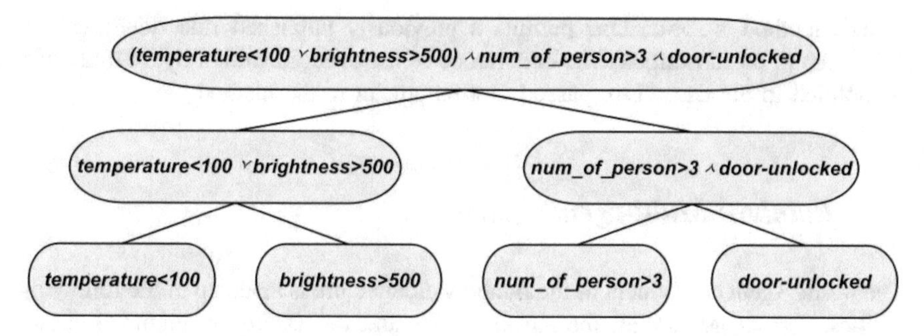

Fig. 2 Example of binary tree of a rule

In the following is described the interface of Gateway that will be used by the overlying layer:

```
interface GatewayInterface {
   void resourceNaming(String name, VOId voId, VOFunctionId
       functionId, VOFunctionParams params);
   VOResult check(String name);
   VOResult check(String name, VOFunctionParams params);
   VOResult acting(String name);
   VOResult acting(String name, VOFunctionParams params);
   void setRule(Rule rule, String idRule);
   void subscribe(String idRule, EventHandler handler);
}
```

The method `resourceNaming` assigns an identification `name` to a given resource supplied by a given VO. A resource is a specific instance of a *functionality* of a VO refined by some *parameters*. In other word, a resource is the above-mentioned triplet: $[VOId, VOFunctionId, Params]$. The name assigned to a resource via `resourceNaming` can be used in the other methods in order to simply identify the resource. Furthermore, the identification `name` of a resource is useful to compose the rules in a more human-readable fashion.

The method `check` reads the current value of the resource identified by name whereas `acting` triggers the actuation operation upon the resource identified by `name`. Both `check` and `acting` methods are of two kinds: the first take only `name` as parameter and refers to the resource as it is previously defined in `resourceNaming`; the second kind, instead, permits the parameters of the referred resource to be refined dynamically.

The method `setRule` permits a complex rule to be published (e.g. *(temperature <100 and brightness >500) or number_of_ person >3 or door_unlocked*) and to assigns an id (i.e. `idRule`) useful for subscribing the rule afterwards.

The method `subscribe` permits a previously published rule (identified by `idRule`) to be subscribed. The occurrence of the event identified by `idRule`will be notified to the `handler` passed as a parameter to the method.

3.2 Rainbow Multi-agent System

The Multi Agent component of the rainbow architecture is made up of the following entities: *Agents*, *Messages*, the *Agent Server* and the *Deployer*. Figure 3 shows these entities and how they interact among themselves and with the *Gateway*.

The *Agent Server* is the container for the execution of agents. It offers functionalities concerning the life cycle of the agents as well as functionalities for agents' communication. Agent servers are arranged in a peer-to-peer fashion where each agent server hosts a certain number of agents and permits them to execute and interact transparently among themselves. In other words, when an agent requests the execution of a functionality, its host agent server is in charge of redirecting transparently the request to the suitable agent server. In the following are listed the main functionalities each agent server exposes:

`SEND_MSG`. Through this functionality, the communication between agents is performed. The Agent Server is responsible for correctly delivering messages from the sender agent to the receiver one. If the sender and the receiver do not belong to the same agent server, the message is forwarded to the suitable "peer" agent server which is, in turn, engaged finally to deliver the message. The latter mechanism is showed in Fig. 4a.
`ADD_AGENT`. It instances an agent to an agent server. Rainbow Multi Agent system is designed to permit agents to be dynamically loaded to the agent server they have to belong to. As in SEND_MSG operation, agent servers are in charge for exchanging information among themselves in order to guarantee the ADD_AGENT request to be delivered to the correct agent server. This mechanism is shown in Fig. 4b. The latter figure also shows how the code is dynamically loaded exploiting *class repository server*. More in detail, when an ADD_AGENT request reaches the

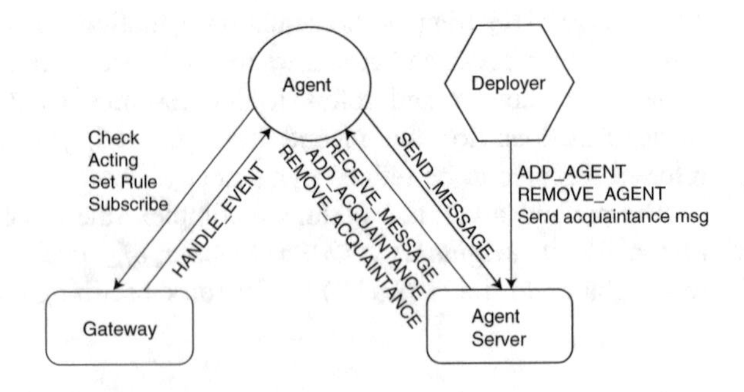

Fig. 3 Rainbow multi-agent entities

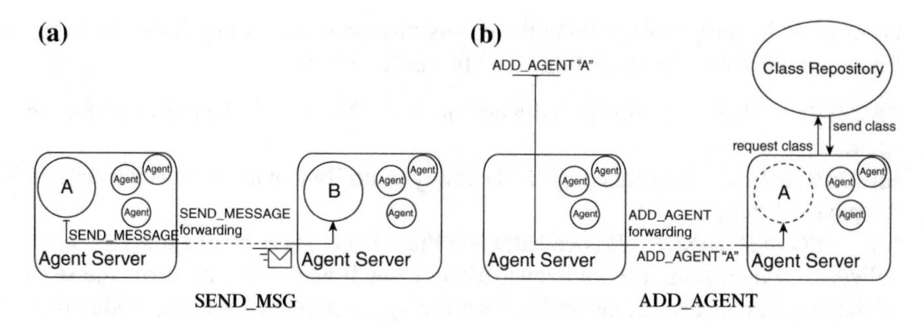

Fig. 4 Forwarding mechanism. **a** [SEND_MSG], **b** [ADD_AGENT]

suitable agent server, if the agent code is not already available, the agent server automatically downloads it from a class repository.
`REMOVE_AGENT`. It removes an instance of an agent hosted by an agent server. This operation also exploits the "forwarding" mechanism described above.

A *Message* is the atomic element of communication between agents. It carries an application specific content together with information about the sender agent and the receiver one.

Our architecture provides for specific kinds of message, that are the *acquaintance messages*. Those messages are used for establishing an acquaintance relationship among agents. The acquaintance message carries information about the location of a given agent (i.e. location of hosting agent server). The agent who receives the acquaintance message will use this information when it needs to send messages toward that destination. This kind of mechanism ensures agent behaviour to be completely independent w.r.t. the locations of agents it has to collaborate with.

For instance, let's consider that an agent is a computing node interconnected with others by means of a ring network. Each agent, therefore, can only interact with its previous agent nodes and its next one. Whenever further nodes must be connected to the ring network, only the acquaintance relationships have to be updated. In other words, a third entity can establish dynamically those acquaintance relationships without resorting to modifying, re-building or restarting any agent.

In rainbow architecture the entity which is in charge of sending acquaintance messages in order to establish the acquaintance network is called *Deployer*. Deployer could be an external process as well as an agent, it can run during the configuration phase as well as during application execution. The Deployer concept will be described in details in Sect. 3.2.1.

An *Agent* is an autonomous entity which executes its own behaviour interacting with other agents via Agent Server. In addition, each agent can interact with the physical part exploiting functionalities exposed by the Gateway (i.e. using the Virtual Object abstraction).

The functionalities of an agent are exposed to its own Agent Server and Gateway. As said before, Agent Servers are in charge of the "forwarding" mechanism that eventually ends with the calling of these functionalities, while the

Gateway is in charge of notifying the events that occur in the physical part. In the following are listed the main functionalities of an agent:

RECEIVE_MESSAGE. It is called when there is a Message to be delivered for the agent.
HANDLE_EVENT. It is called by the Gateway to notify that an event is occurred in the physical part.
ADD_ACQUAINTANCE. It is called when there is an acquaintance message to be delivered to the agent. The implementation of this functionality concerns the store of the acquaintance relationship between the agent itself and the agent identified inside the message.
REMOVE_ACQUAINTANCE. It is called for removing a previously stored acquaintance relationship.

The specific behaviour of an Agent is realized through the implementation of RECEIVE_MESSAGE and HANDLE_EVENT functionalities.

3.2.1 Dynamic Deployment and Roles

The deployment of the agents as well as the configuration of the acquaintance relationships and the start-up of the application are all actions performed by the so-called *Deployer*. An external process or even an agent can act as a Deployer. The deployment phase is typically executed just before the application can start properly; however, it is possible to act as Deployer even during application execution in order to update the configuration dynamically for hosting new features or adapting to foreseen and unforeseen changes in the environment. Deployer can be implemented centrally or in a distributed way. Basically, who acts as a Deployer operates using the ADD_AGENT functionality for deploying a new instance of an agent into an agent server, REMOVE_AGENT for removing a running agent from an agent server. Furthermore, Deployer is responsible for sending acquaintance messages that eventually end with calls to ADD_ACQUAINTANCE or REMOVE_ACQUAINTANCE on the specific agents. Finally, Deployer is also in charge of sending suitable "start" messages using SEND_MSG in order to start the application properly.

The acquaintance relationship is formally defined by a triplet: [A, B, R] where A and B are the agents involved in the relationship and R is a *Role* label. The triplet above means that agent A knows agent B and that B has the role R as acquaintance of A. During the execution, an agent exploits the Roles of its acquaintances to discriminate about how to interact with them.

As an instance, let's consider that each agent represents a physical person in a town. The relationship between two agents could have roles of neighbourhood and/or friendship. A deployer is in charge of configuring those relationships during the initial phase. In addition, as soon as a person changes home or starts a new friendship, the deployer has to re-arrange relationships dynamically among agents through sending acquaintance messages. During the execution of that system, each agent will use roles of neighbourhood and friendship to discriminate how to interact with other agents. For instance he/it can exchange information about its district with its neighbours while it invites its friends to a party.

4 Application Examples

In this section we introduced two examples of use the Rainbow architecture. The first one aims to show our architecture from a practical perspective in order to understand and better figure out all the system details. The second example is useful to understand how Rainbow can host suitable swarm intelligence strategies in order to realize CPS applications owning properties such as adaptivity, fault tolerance, self-reconfiguration, etc.

4.1 Floor Control Example

In this example we show an application for monitoring and controlling a floor of a building hosting offices. Each floor contains a certain number of rooms.

Figure 5 shows how a generic floor could be. In general, each room contains: doors, desks, chairs and adjustable brightness lights.

Each room is instrumented by some sensors and actuators listed below.

Sensors:

- sensors that detect the opening and closing of doors;
- sensors that detect when a person enters or leaves a room;

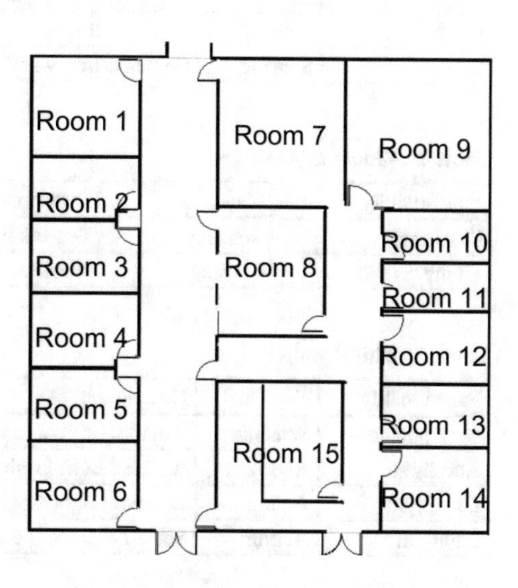

Fig. 5 Floor topology

- proximity sensors detecting presence of the people in each zone of a room;
- a weight sensor for each chair in order to detect if the chair is currently used.

Actuators:

- adjustable brightness lights for all zones of a room;
- a display on each desk.

The use of the above described devices, for example, permits adjusting lights on the basis of people movements, writing informational messages on displays and so on.

4.1.1 Integration in Rainbow Using Virtual Objects

In order to develop the controlling part in a object-oriented fashion, it is required to integrate the above described physical things with Rainbow middleware defining the suitable Virtual Object (VO). Each VO abstracts and wraps a certain number of sensors as well as actuators. For the sake of simplicity, in this example we chose to design VOs in a human-readable fashion: *virtual desk*, *virtual chair*, *virtual door* and *virtual wall*.

The functionalities exposed by these VOs are listed in Tables 1, 2, 3 and 4. It is worth to note that each functionality of the virtual wall is parametric: the *zone* parameter specifies which area of the room is referred.

Table 1 Virtual door

Functionality	Type	Description
Lock	Sensing	Boolean, true if the door is closed
Unlock	Sensing	Boolean, true if the door is open
Entry	Sensing	Boolean, true when a person enter the room through the door
Exit	Sensing	Boolean, true when a person exit the room through the door

Table 2 Virtual chair

Functionality	Type	Description
Proximity	Sensing	Detects people near the chair
Sitting	Sensing	Boolean: true when someone sits on the chair

Table 3 Virtual wall

Functionality	Type	Description
Near people	Sensing	Number of people in the zone (supplied by parameter)
Add light	Acting	Increase light brightness in the zone (supplied by parameter)
Less_light	Acting	Decrease light brightness in the zone (supplied by parameter)
Light_off	Acting	Set off light in the zone (supplied by parameter)

Table 4 Virtual desk

Functionality	Type	Description
Proximity	Sensing	Detects people near the desk
Display	Acting	Show a message supplied by parameter on the display

Each VO is located on the same computing node where the sensors and actuators that VO encloses are connected to. A computational node can generally host VOs that may refer to more than one room. Assuming than we have only three computational nodes available to monitor and control the whole floor, we can assign rooms to nodes as in Fig. 6.

4.1.2 Multi-agent Floor Application

The application is designed for managing the floor and its rooms. For each room a energy-saving light-management is developed which considers people presence for suitably adjusting the brightness of the various zones of a room. This control management will also consider if the chairs are utilized or not in order to better adjust the lights. In addition, it permits a message to be displayed on a certain desk when needed. All those features are implemented in the *RoomAgent*. The code inside the RoomAgent is a typical object-oriented code where VOs are exploited as simple objects. The code is omitted in this paper for sake of brevity.

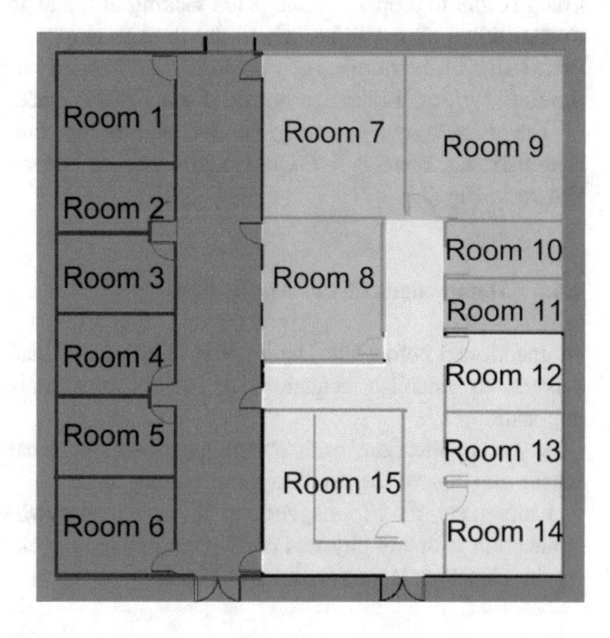

Fig. 6 Rooms assignment to computational node. Each different color identify a different node

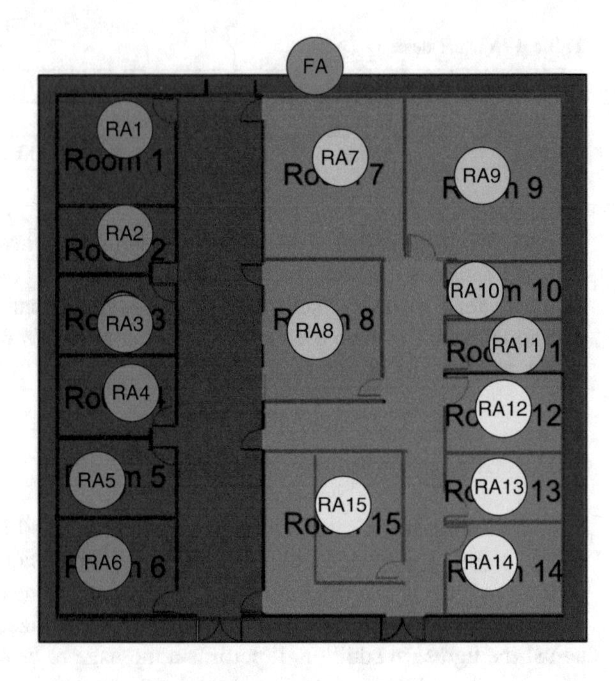

Fig. 7 Logical distribution of agents in the floor

Besides this room-wise features, the application is also designed for addressing issues concerning the entire floor (i.e. where more than one room is involved). For instance, it could be useful to know how many people are in the floor at a given time in order to properly manage the locking of the main door of the floor as well as to shut down all the lights where the floor is empty. In this example, instead, the knowledge of the number of people is used to notify a person when he is alone in the floor writing a message on the display of his desk.

The *FloorAgent* is designed for addressing the above mentioned issues. Summarizing, there is a *RoomAgent* per room and a unique *FloorAgent* as it is shown in Fig. 7.

4.1.3 Deployment of the Application

As mentioned before the *Deployer* is in charge to load the agents upon the agent servers, to establish acquaintance relationships among them and to start the application.

In our application, each *RoomAgent* must be located in the computing node where the VOs of the relative room belong to.

Conversely, the *FloorAgent* can be located everywhere in the system (it has not connection with any physical part), even in a remote cloud node. The process made by the *Deployer* is summarized in Fig. 8.

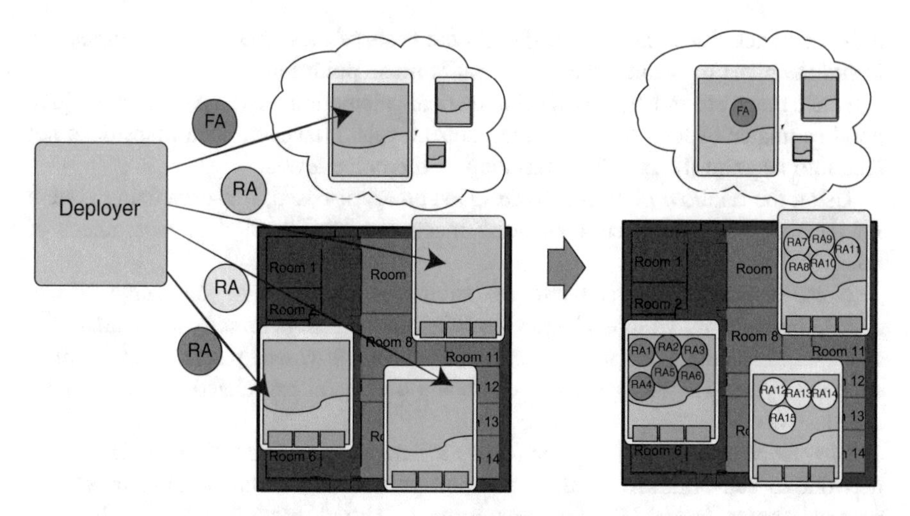

Fig. 8 Deployment of the agents and their physical distribution on the computing nodes

4.1.4 Agent Interaction and Acquaintance Relationships

After loading each agent in the proper location, the *Deployer* sends acquaintance messages to each *RoomAgent* in order to let them know the *FloorAgent*. Afterwards, each agent sends an acquaintance message to the *FloorAgent* in order to be known by it. This is an example of an agent that acts as *Deployer*. Once the deployment phase is completed, the application execution can start. When a person leaves a room, *RoomAgent* will be notified by the gateway and, consequently, will send a message carrying the number of people currently inside the room to the *FloorAgent*. The latter will update its people counter on receiving such a message. When it verifies that there is only one person in the floor, it will send a message to the relative *RoomAgent* that, in turn, will write a message on the desk display.

4.2 Noise Pollution Mapping

Many environments, such as airports, road works, factories, construction sites, and other environments producing loud noises, require effective noise pollution monitoring systems. Noise pollution is a common environmental problem that affects people's health by increasing the risk of hypertension, ischemic heart disease, hearing loss, and sleep disorders, which also influence human productivity and behavior [19]. For this reason the European Community passed the directive 2002/49/EC [20], which declares noise protection as one necessary objective to achieve a high level of health and environmental conservation. The directive imposes several actions to be made upon member states, including the mapping of noise in larger cities via noise maps. On the basis of these maps, the countries can formulate plans to counter the threat that is noise pollution.

Noise maps are mostly based on numerical calculations that have shown to give good estimates of long term averaged noise levels. However, such maps does not take into account the real-time variation of the noise levels.

Using the rainbow platform we designed an agent-based, self-organizing system for the real-time construction of noise maps and identification of the sources of noise.

Noise sensors are spread into the environment, linked to the computational nodes, and suitably wrapped inside the VOs. Each agent is directly associated with a VO representing a noise sensor. During the deployment phase, each agent is supplied by the knowledge of its neighbours (i.e. agent associated with a spatially near sensor).

We use a simple self-organizing algorithm, proposed by [21], to let sensor network to self-organize itself in a region partitioning based on similar sensing patterns (*noise levels*). Regions can grow or shrink according to the dynamic variation of noise levels. Organization in regions occurs by creating an overlay network made by agents connected by virtual weighted links. Agents belonging to the same region will have strong links, while agents belonging to different regions will have weak (or null) links.

In the following the details of the algorithm. Let s_i and s_j be two neighbour sensor agents. Let $n(s_i)$ and $n(s_j)$ the values of noise sensed by s_i and s_j, respectively. Let us assume that a distance function D can be defined for couples of v values. Region formation is then based on iteratively computing the value of a logical link $l(s_i, s_j)$ for each and every agent of the system as in following update_link procedure:

Update_link:

$$
\begin{aligned}
&if(D(n(s_i), n(s_j)) < T\{\\
&\quad l(s_i, s_j) = min(l(s_i, s_j) + \Delta, 1)\\
&\}else\{\\
&\quad l(s_i, s_j) = max(l(s_i, s_j) - \Delta, 0)\\
&\}
\end{aligned}
$$

where T is a threshold that determines whether the measured values are close enough for $l(s_i, s_j)$ to be re-enforced or, otherwise, weakened; and Δ is a value affecting the reactivity of the algorithm in updating link. Based on the above algorithm, it is rather clear that if $D(n(s_i), n(s_j))$ is lower than threshold T, $l(s_i, s_j)$ will rapidly converge to 1. Otherwise it will move towards 0. Transitively, two nodes s_h and s_k are defined in the same region if and only if there is a chain of agents such that each pair of neighbours in the chain are in the same region. From the Rainbow perspective, region information is stored adding/removing new acquaintance relationships among agents.

In order to properly map the noise pollution, it is necessary that each and every agent within a region is locally provided with information related to the overall status of the region. To this end, it is possible to integrate forms of diffusive gossip-based aggregation [22] within the described general scheme. The algorithm requires that the agents periodically exchange information with their neighbors about some local value, locally aggregate the value according to some aggregation function (e.g., maximum, minimum, average, etc.), and further exchange in the subsequent step the aggregated value.

5 Conclusions

In this paper we introduced Rainbow, an architecture that permits an easy development of large-scale cyber-physical applications. The novelty of Rainbow is that it relies on the adoption of a distributed multi-agent layer on top of the physical part that is, in turn, wrapped in suitable Virtual Objects. Rainbow aims to hide heterogeneity, cope with complexity and real-time issues. In the future, new intelligent, adaptive and decentralized algorithms will be explored for developing large-scale cyber-physical applications using Rainbow, such as those related to smart cities, power grid, water networks and so on. Furthermore, a well-established interface for the cloud part of the architecture will be defined.

Acknowledgments This work has been partially supported by RES-NOVAE—"Buildings, roads, networks, new virtuous targets for the Environment and Energy" project, funded by the Italian Government (PON 04a2_E).

References

1. Koubaa, A., Andersson, B.: A vision of cyber-physical internet. In: Proceedings of the Workshop of Real-Time Networks (RTN 2009), Satellite Workshop to (ECRTS 2009)
2. Lee, A.: Cyber physical systems: design challenges. In: Proceedings of the 2008 11th IEEE Symposium on Object Oriented Real-Time Distributed Computing, IEEE Computer Society Washington, DC, USA (2008)
3. Sanislav, T., Miclea, L.: Cyber-physical systems—concept, challenges and research areas. Control Eng. Appl. Inf. **14**(2), 28–33 (2012)
4. Shi, J., Wan, J., Yun, H., Suo, H.: A survey of cyber-physical systems. In: Proceedings of the International Conference on Wireless Communications and Signal Processing, Nanjing, China, 9–11 Nov 2011
5. Abdelzaher, T.: Towards an architecture for distributed cyber-physical systems. In: Proceedings of NSF Workshop on Cyber-Physical Systems, Austin, TX (2006)
6. Bicocchi, N., Mamei, M., Zambonelli, F.: Self-organizing virtual macro sensors. ACM Trans. Auton. Adapt. Syst. (TAAS) **7**(1) (2012)
7. Fortino, G., Guerrieri, A., Lacopo, M., Lucia, M., Russo, W.: An agent-based middleware for cooperating smart objects. In: Highlights on Practical Applications of Agents and Multi-Agent Systems, Communications in Comp. and Inform. Science (CCIS), vol. 365, pp. 387–398. Springer (2013)

8. Leito, P.: Towards self-organized service-oriented multi-agent systems. In: Studies in Computational Intelligence, vol. 472. Springer (2013)
9. Lin, J., Sedigh, S., Miller, A.: Modeling cyber-physical systems with semantic agents. Computer Software and Application Conference Workshops (COMPSACW), IEEE (2010)
10. Bonabeau, E., Dorigo, M., Theraulaz, G.: Swarm Intelligence: From Natural to Artificial Systems. Oxford University Press, Santa Fe Institute Studies in the Sciences of Complexity, New York, NY, ISBN: 0-19-513159-2 (1999)
11. Kennedy, J., Eberhart, R.C.: Swarm Intelligence. Morgan Kaufmann publishers (2001)
12. Foundation for Innovation in Cyber-Physical Systems, National Institute of Standard and Technology (NIST) (2013)
13. RaspBerry online. http://www.raspberrypi.org/
14. Dillon, T.S., Zhuge, H., Wu, C., Singh, J., Chang, E.: Web-of-things framework for cyber–physical systems. Concurrency and Computation: Practice and Experience. **23**(9), 905–923 (2011)
15. Hnat, T.W., Sookor, T.I., Hooimeijer, P., Weimwe, W., Whitehouse, K.: MacroLab, A vector-based macroprogramming framework for cyber-physical systems. In: Proceedings of the ACM Conference on Embedded Networked Sensor Systems, Raleigh, NC, U.S.A. (2008)
16. Luo, L., Abdelzaher, T.F., He, T., Stankovic, J.A..: Envirosuite: An environmentally immersive programming framework for sensor networks. ACM Transactions on Embedded Computing Systems (TECS). **5**(3), 543–576 (2006)
17. Newton, R., Morrisett, G., Welsh, M.: The regiment macroprogramming system. In: Proceedings of International Conference on Information Processing in Sensor Networks, Cambridge, MA, U.S.A. (2007)
18. Talcott, C.L.: Cyber-physical systems and events. In: Software-Intensive Systems and New Computing Paradigms, 101–115 (2008)
19. Schweizer, I., Brtl, R., Schulz, A., Probst, F., Mhlhuser, M.: NoiseMap—real-time participatory noise maps. ACM SenSys 2011 Second International Workshop on Sensing Applications on Mobile Phones (Eds.) (2011)
20. European Directive.: The Environmental Noise Directive (2002/49/EG). Official Journal of the European Communities, 2002. Available at http://eur-lex.europa.eu/legal-content/EN/TXT/?uri=CELEX:32002L0049 (last visited in November 2015)
21. Bicocchi, N., Mamei, M., Zambonelli, F.: Self-organizing virtual macro sensors. ACM Transactions on Autonomous and Adaptive Systems (TAAS). **7**(1), 2 (2012)
22. Jelasity, M., Montresor, A., Babaoglu, O.: Gossip-based aggregation in large dynamic networks. ACM Trans. Comp. Syst. **23**(3), 219–252 (2005)

Towards Cyberphysical Digital Libraries: Integrating IoT Smart Objects into Digital Libraries

Giancarlo Fortino, Anna Rovella, Wilma Russo and Claudio Savaglio

Abstract Digital libraries are distributed software infrastructures that aim at collecting, managing, preserving, and using digital objects (or resources) for the long term, and providing specialized services on such resources to its users. Service provision should be of measurable quality and performed according to codified policies. Currently, modern digital libraries include a wide range of conventional digital objects: text document, image, audio, video, software, etc. In the emerging domain of the Internet of Things (IoT), cyberphysical smart objects (or simply smart objects) will play a central role in providing new (smart) services to both humans and machines. It is therefore challenging to include smart objects, the newest type of digital objects, into digital libraries as novel first-class objects to be collected, managed, and preserved. However, their inclusion poses critical issues to address and many research challenges to deal with. This paper aims at paving the way towards such a novel inclusion that will enable effective discovery, management and querying of smart objects, so establishing cyberphysical digital libraries. In particular, our approach is based on a metadata model purposely defined to describe all the cyberphysical characteristics (geophysical, functional, and non-functional) of smart objects. The metadata model is then used for a seamless integration of smart objects into digital libraries compliant with the digital library

G. Fortino (✉) · W. Russo · C. Savaglio
DIMES, University of Calabria, via P. Bucci, cubo 41C, 87036 Rende, CS, Italy
e-mail: g.fortino@unical.it

W. Russo
e-mail: w.russo@unical.it

C. Savaglio
e-mail: csavaglio@dimes.unical.it

G. Fortino
CNR—National Research Council of Italy, Institute for High Performance Computing and Networking (ICAR), via P. Bucci, 87036 Rende, CS, Italy

A. Rovella
LISE, University of Calabria, via P. Bucci, cubo 20B, 87036 Rende, CS, Italy
e-mail: anna.rovella@unical.it

A. Guerrieri et al. (eds.), *Management of Cyber Physical Objects in the Future Internet of Things*, Internet of Things,
DOI 10.1007/978-3-319-26869-9_7

reference model proposed by the DL.org community. The proposed approach is also exemplified through a simple yet effective case study.

Keywords Internet of Things · Cyberphysical smart objects · Digital libraries · Metadata

1 Introduction

Digital Libraries (DLs) have undergone a considerable evolution, becoming complex entities, able to manage and preserve different types of digital material [1]. They offer a variety of services that can be pervasive and ubiquitous and can be heterogeneous in characteristics, objectives and functions. Since the 1990s [2] librarians first and researchers belonging to different fields later have elaborated different theories and applications and for this reason the definition of DL presents a polysemy of meanings that reflects different visions and approaches. The concept of DL has therefore evolved [3], moving from a system for the retrieval of static information (primarily books and digitalized textual documents) to a tool useful for the collaboration and interaction between researchers and users, regarding domain-specific topics. Currently, DLs include a wide range of digital objects: text document, image, audio, video, software, etc. [4].

In the emerging domain of the Internet of Things (IoT) [5, 6], a novel type of digital resource is the cyberphysical smart object (SO). An SO is a daily life physical object augmented with sensing/actuation, processing, storing, and networking capabilities, in order to provide a set of physical and digital services to its users (both humans, machines, or digital systems) [7–9]. During their lifecycle, SOs can produce continuous streams of geolocalized and contextual data also related to their use and their surrounding environment. Moreover, SOs may evolve to provide new/different cyberphysical services to their users.

This chapter extends the proposal in [10], presenting an approach for the inclusion of SOs into DLs which would enable effective discovery, querying and management of SOs based on typical DL tools and facilities. To the best of our knowledge, this approach represents the first research effort towards the integration of SOs into DLs in order to establish cyberphysical DLs. In particular, the approach is based on a well-defined metadata model for SOs able to describe all the cyberphysical characteristics (geophysical, functional, and non-functional) of SOs. The SO metadata model is used for the inclusion of SOs into DLs compliant with the digital library reference model [11] proposed by the DL.org community [12].

The remainder of this paper is organized as follows: Sect. 2 discusses work related to operational and non-operational metadata models for SOs. In Sect. 3, we define the proposed SO metadata model in detail whereas in Sect. 4, a case study is presented to exemplify all concepts of the SO metadata model. Section 5 presents a brief overview about the main concepts related to the DL universe and to the Digital Library Reference Model (DLRM), while Sect. 6 provides an articulated discussion about the inclusion of SOs into DLs. Finally, conclusions are drawn and future research efforts are anticipated.

2 Related Work

SOs will represent the basic intelligent entities constituting the future IoT and its related IoT applications [5]. There is therefore a need to define a reference metadata model for SOs that can facilitate their management from different perspectives (e.g. internal status, provided services, distributed discovery, and interaction with the physical world, the user and other systems) and their inclusion in highly dynamic and complex ecosystems (e.g. IoT, Internet of the Future, and next-generation DLs).

In the literature, many works are available, in which the SO definition and the consequent inclusion in existing architectures is very differently argued. Among these, it is possible to recognize operational and non-operational SO metadata models, as Table 1 summarizes.

Models proposed in [5] and [13] belong to the non-operational models. In [13], an SO classification according to the concepts of creator and purpose is defined. In particular, the creator can be either an individual creating SOs for a personal purpose (e.g. personal use) or an industrial company that creates SOs for business. The former SOs are called *self-made* whereas the latter ones are named *ready-made*.

Table 1 Comparison among non operational and operational SO metamodels

	Descriptive features	Operative features
Non operational SO metamodel		
[13]	SO classified according to creator (*self-made, ready-made*) and to purpose (*specific, open-ended*)	N/A
[5]	SO classified into design dimensions of *activity-awareness, policy-awareness* and *process-awareness*	N/A
Operational SO metamodel		
[14]	SO described by exploiting the SO meta-information (name, vendor, etc.) contained in the *SODD* document	SO managed by exploiting the *PDD* document, which contains information about SO devices and functionalities
[15]	SO basic information is contained in a queryable *Resource Registry*	SO relationships with its physical counterpart (*Digital Proxy*) and its *User* elicited by the SO model
[16, 17]	SO basic information (*type, location, etc.*) spread among the SO model categories	SO indexing, discovery and dynamic selection realized by exploiting SO model categories such as *services, devices*, etc.

The purpose of an SO may be to play a role in a specific application/system or to be reused in a wide range of different applications. The former is defined *specific*, while the latter *open-ended*. However, such a classification considers only two dimensions (creator and purpose) that are not related to the cyberphysical characteristics of the SOs. Thus, such classification cannot be used in an operational way within an IoT system. In [5] authors classify SOs in *activity-aware*, *policy-aware*, and *process-aware*. Each SO type is characterized by three design dimensions: (i) awareness, which is the ability of SOs to understand (environmental or human) events of the SO surrounding context; (ii) representation, which refers to the programming model of the SO; and (iii) interaction, which defines the communication with users. Such classification is oriented to the design of SOs within an application domain and can be usefully exploited during IoT systems development. However, such contribution is not operational as it can only be used to classify SOs according to design dimensions.

We are indeed interested in operational classifications that are the base to build up SO discovery services and management systems. In [14] the operational SO classification is based on two documents: Smart Object Description Document (*SODD*) and Profile Description Document (*PDD*). *SODD* contains the meta-information of the SO: name, vendor, and list of profiles. *PDD* specifies a profile that can be either a detector or an actuator. A detector contains information about a specific sensing device according to the Sensor Modeling Language (SML), whereas an actuator is modeled through the Actuator Modeling Language (AML). The proposed classification is specific to the SO implementation and management supported by the FedNet middleware [14]. In [15] two main concerns are addressed through a conceptual technological agnostic model: (i) the interactions between the *User* (human or not) and the SO, (ii) the synergy between the Physical Entity and the Digital counterpart. In this direction the *Digital Proxy*—which is the representation of a given set of aspects (or properties) of the Physical Entity—plays the crucial roles of SO identifier and bridge between the real and the virtual world. In fact, the functionalities of sensing and actuation are delegated to the Devices, which therefore realize the effective interaction with the physical reality. SO basic information is instead contained in a queryable *Resource Registry*. Finally, concepts such as aggregation between SOs (which can be logically grouped in a structured, often hierarchical way) and the relationship between Services and Resources provide flexible guidelines for an SO modeling that ensures interoperability with and openness to functional and technological developments not entirely predictable. In [16], a metadata model to represent functional and non-functional characteristics of SOs in a structured way is proposed. The metadata model is divided into four main categories: *type*, *device*, *services*, and *location*. The type is the SO type (e.g. smart pen, smart table, etc.). The device defines the hw/sw characteristics of the SO device. Services contain the list of services provided by the SO; in particular, a SO service can have one or more operations implementing it. The location represents the position of the SO. This metadata model, which is more general than the one proposed in [15], has been extended and re-organized in [17] and it is currently implemented in a discovery framework (named SmartSearch) for SO indexing, discovery and dynamic selection [16].

3 A Metadata Model for CyberPhysical Smart Objects

The proposed metadata model is an extension of the model proposed in [16, 17] and also borrows some concepts from the other models discussed in Sect. 2. The metadata model is portrayed in Fig. 1 according to the UML class diagram formalism. In particular, the proposed model defines a set of metadata categories that can characterize an SO in any application domain of interest (e.g. Smart Cities, Smart Factories, Smart Home, Smart Grid, Smart Building, etc.). The metadata represent the SOs static parameters, while the related dynamic parameters can be retrieved through operations associated to the available services or from the smart object status (usually through basic SO status services). In our metadata model, an SO, which could aggregate other SOs according to the aggregation relationship, is a composition of the following main metadata categories:

- **Status**: is characterized by a list of variables, given as pairs ⟨name, value⟩, that capture the SO state.
- **FingerPrint**: contains the following basic and distinctive SO information:

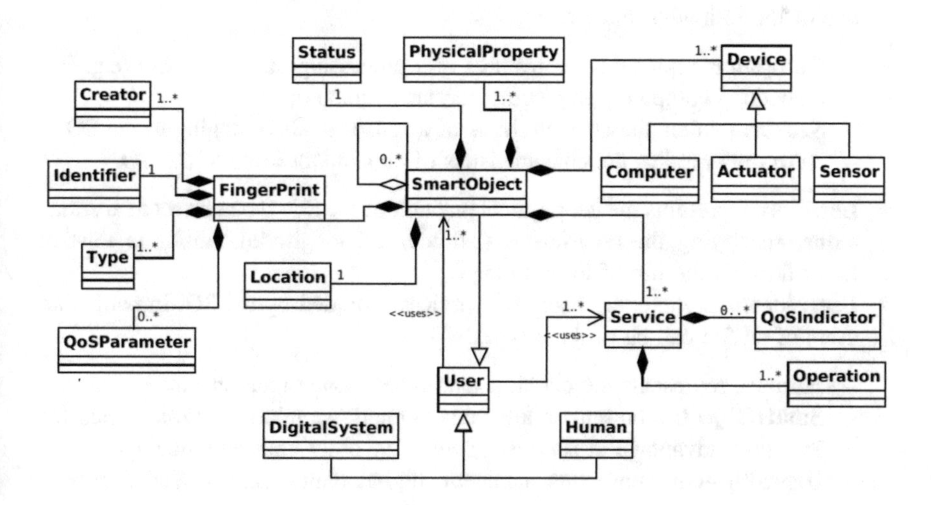

Fig. 1 The SO metadata model

- **Identifier**: represents the identifier (or Id) of the SO, which allows its unique identification within the IoT or an IoT subsystem;
- **Creator**: represents the SO creator, which can be either an individual creating the SO for personal use, an industrial company that creates it for business, or an academic research laboratory implementing it for research purposes;
- **Type**: represents a primary type of SO (e.g. a smart pen, a smart chair, a smart office). Moreover, secondary SO types can also be given that contain, for instance, information about the SO design classification as proposed in [5];
- **QoSParameter**: defines a QoS parameter associated to the SO. Different QoS parameters may be defined such as trustness, reliability, availability, etc.

- **PhysicalProperty**: represents a physical property of the original object without any hardware augmentation and embedded smartness.
- **Service**: models a digital service provided by the SO. A service has a name, a description, the type (sensing, actuation, SO status), input parameters, and the return (primitive or complex) parameter type. Each Service is characterized by one or more **Operation**s that implement the service itself and by zero or more **QoSIndicator**s whose associated values are provided. In particular, an **Operation**, which defines the individual operation that may be invoked on a service, is equipped with a set of input parameters necessary for its invocation, the return (primitive or complex) parameter type and a description.
- **Device**: defines the hardware and software characteristics of a device that allows to augment the physical object and make it smart. Device can be specialized into one of the following three categories:
  - **Computer**: represents the features of a processing unit of the SO (e.g. PC, embedded computer, plug computer, smart-phone);
  - **Sensor**: models the characteristics of a sensor node belonging to the SO;
  - **Actuator**: models the characteristics of an actuator node of the SO.
- **Location**: represents the geophysical position of the SO. It can be set in absolute terms, specifying the coordinates (latitude and longitude), and/or in relative terms through the use of location tags.
- **User**: identifies the entity using the services provided by the SO. In particular, users of an SO can be of three types:
  - **Human**: represents the classical man-object usage relationship;
  - **SmartObject**: represents a less conventional use relationship, in which the SOs take advantage of services exposed by other ones and vice versa;
  - **DigitalSystem**: represents a generic digital entity, like a Web Server, a software agent, a robot or even more complex systems.

4 A Case Study: Smart Office

The objective of this section is to show the instantiation of the SO metadata model introduced in Sect. 3 with respect to a case study referring to the SO "SmartOffice" defined in [16]. The SmartOffice, on the basis of the information gathered and a set of inference rules, supports office users during their daily working activity by providing suggestions (e.g. warnings that it would be appropriate to take a break after a long session of work by sitting, or indicating how to adjust the screen brightness based on the room luminosity showing such information on the screen closest to the user) and performing smart actuations (e.g. turning the lights and/or the projector off while not used in order to avoid energy wastage). In particular, the SmartOffice provides services obtained by the cooperation of multiple heterogeneous (but at same time independent) SOs located in the office area or worn by the office users. The SmartOffice with its aggregates SOs and its devices are illustrated in Fig. 2, while the SmartOffice services are summarized in Table 2.

The SmartOffice model, which is obtained by instantiating the SO metadata model, can be partitioned into six parts (see Fig. 3):

- *Smart Object Core*: SmartOffice contains the current status information about the SmartOffice itself (temperature, humidity, presence, and light variables). Moreover, the UML aggregation and *uses* relationships underline respectively that Smart Office aggregates and uses three SOs and interacts with different users.

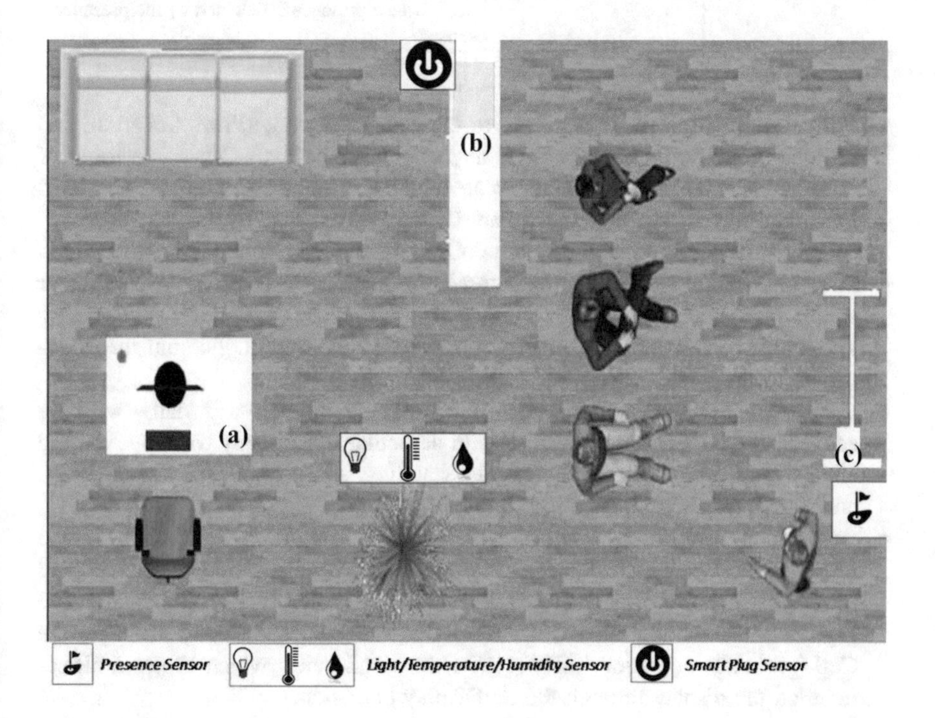

Fig. 2 SmartOffice and its aggregated SOs: **a** SmartDesk, **b** SmartProjector, **c** Smart Whiteboard

Table 2 Smart Office and aggregated SOs services

Reference SO	Service	Description
SmartOffice	PresenceService	Detects the presence of people inside the office, exploiting the information provided by the aggregated SOs and by their presence sensors
	LightService	Informs if the office lights are switched on/off, exploiting the information provided by office light sensor
	VisualizationService	Shows notifications on the available displays, such as the monitor of the user's laptop or the user's smartphone
SmartDesk	PresenceService	Detects whether or not the user is at desk, exploiting the information provided by the presence sensor placed on the desk itself
	VisualizationService	Displays messages on the laptop monitor, providing notifications about the current office status (temperature, humidity, etc.) or warnings (energy wasting, too long sitting user period, etc.)
SmartProjector	ElectricityService	Queries the projector status and controls its energy consumption, exploiting data provided by the smart plug sensor
	VisualizationService	Handles the projection functionality
SmartWhiteboard	PresenceService	Recognizes the whiteboard exploitation by an user, whose presence is detected by the presence sensor

- *BasicFeatures*: information related to the categories FingerPrint, Location, and PhysicalProperty, finds place here. In particular, the Smart Office is a room of dimensions 500 × 700 × 230 cm, located at DIMES-Unical, Cube 41c, 4th floor. Its fingerprint shows that the Smart Office was created by the "SenSysCal" company, is identified with the name "Office1" and has a trustness score of 0.95.
- *Devices*: a light/humidity/temperature sensor (identified as Sensor1) gather simple but useful in-office environmental information. Other devices belonging to aggregated SOs are not directly managed by the SmartOffice but they contribute to the SmartOffice's services realization.
- *Users*: SmartOffice supports the office user Antonio in his daily working activity, and the SO SmartBody [18]. In particular, SmartBody worn by Antonio allows to recognize user activities like standing, sitting, walking, laying down, and to deliver such information to the SmartOffice.
- *Services*: SmartOffice provides different services (see Table 2): (i) the PresenceService, which detects the presence of people inside the office, and provides such information through the GetPresence operation; (ii) the LightService, which informs if the lights are switched on/off through the GetLightStatus operation; (iii) the VisualizationService, which shows notifications on the display through the SetDisplay operation.

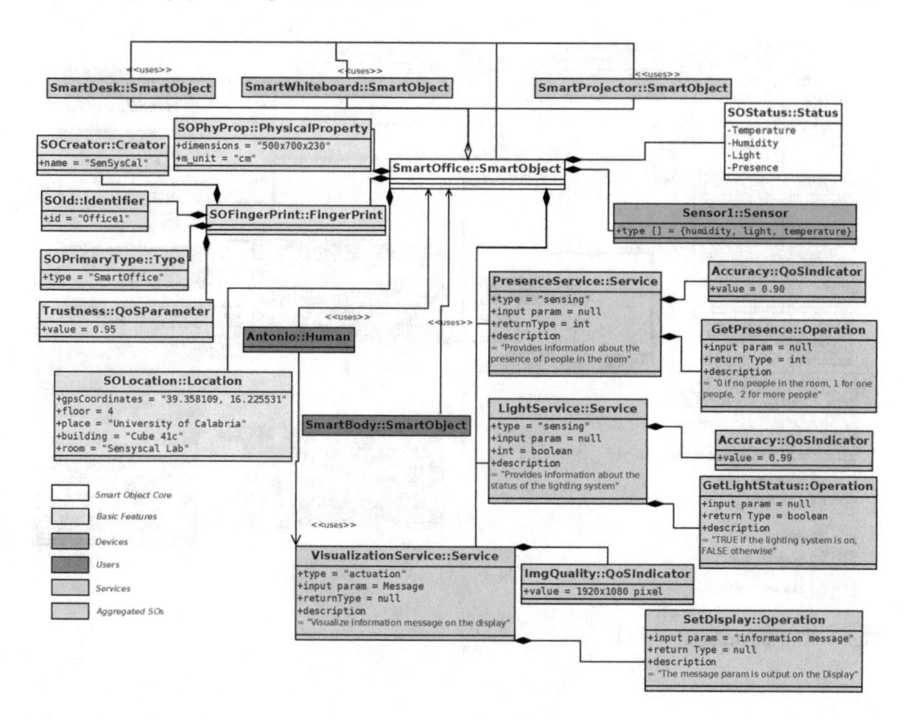

Fig. 3 The SmartOffice model

- *Aggregated SOs*: SmartOffice aggregates the following SOs to implement the SmartOffice services (see Table 2). The single SOs are loosely coupled with the SmartOffice, since they are independent in principle but for the specific application scenario their reference SO is the SmartOffice. In particular:

  – the SmartDesk, whose metadata model is shown in Fig. 4, is equipped with a presence sensor and it provides a (i) PresenceService able to detect whether or not the user is at desk (and for such feature, it's secondary type is "activity aware"); (ii) Visualization Service for displaying messages on the monitor;
  – the SmartProjector, whose metadata model is shown in Fig. 5, is equipped with a smart plug and it provides (i) ElectricityService to query the projector status and to control its energy consumption; (ii) VisualizationService to manage the projection functionality;
  – the SmartWhiteboard, whose metadata model is shown in Fig. 6, is equipped with a presence sensor and it provides a PresenceService able to recognize its exploitation by a user.

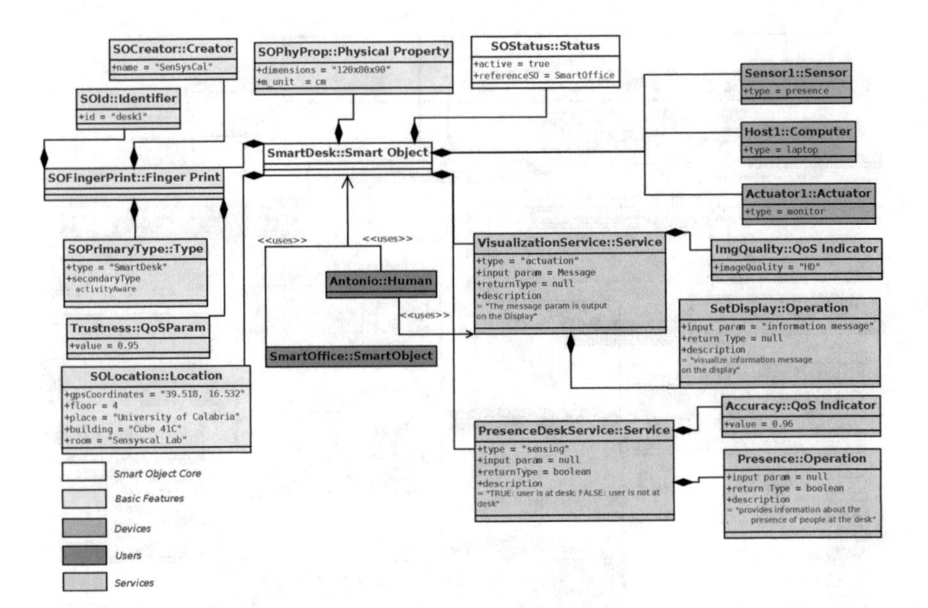

Fig. 4 The SmartDesk model

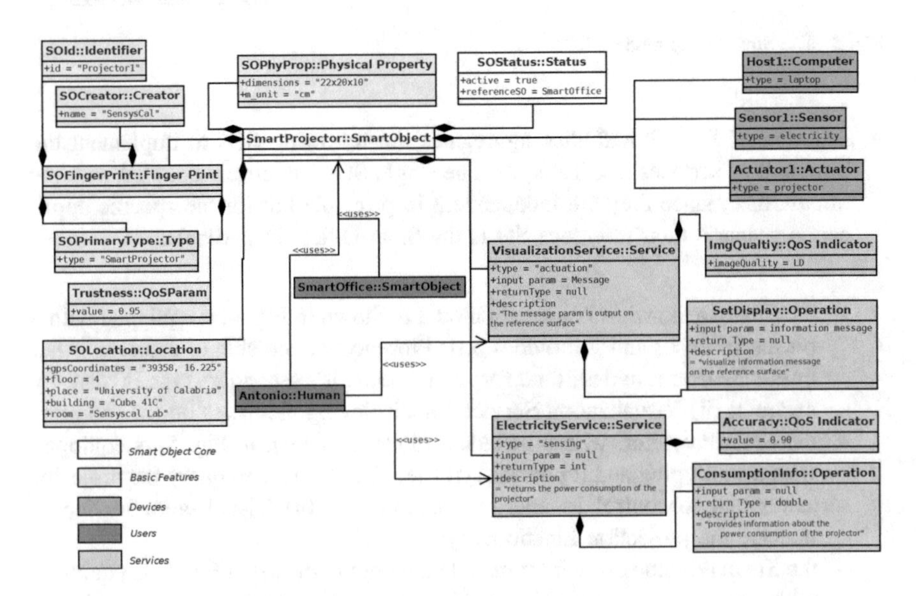

Fig. 5 The SmartProjector model

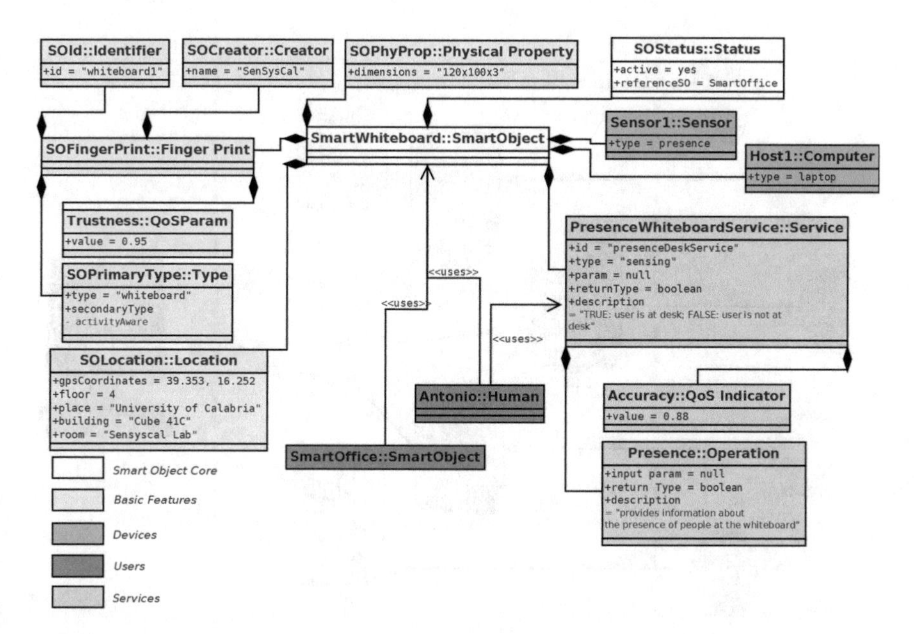

Fig. 6 The SmartWhiteboard model

5 Digital Libraries and the Digital Library Reference Model

Digital Libraries are distributed software infrastructures [1] that comprehensively collect, manage and preserve rich digital content, namely entities in which one or more content files and their corresponding metadata are united, physically and/or virtually, by means of digital wrappers. On the basis of the latter definition, heterogeneous kinds of material may find place into DLs, like textual or multimedia documents, 3-D objects, as well as datasets, databases and data streams coming from sensors, mobile devices or distributed computing. Such content are accessible through specialized functionalities and protocols [19] for the benefits of targeted communities of users, according to specific policies aimed at preserving the quality of content and, therefore, of the library itself.

Although the DL concept represents an abstract system in which both virtual and physical components cooperate to provide services, it presupposes concrete software systems which comprehensively support the DL's lifecycle. In particular, as shown in Fig. 7, the DL universe is structured as a three-tier framework [11], in which the DL's services are provided by handful data management software named DLMSs (Digital Library Management Systems) running atop distributed software system, the DLS (Digital Library System). The transparent interaction and seamless integration of multiple DLSs support the creation of a full-fledged DL, intended as

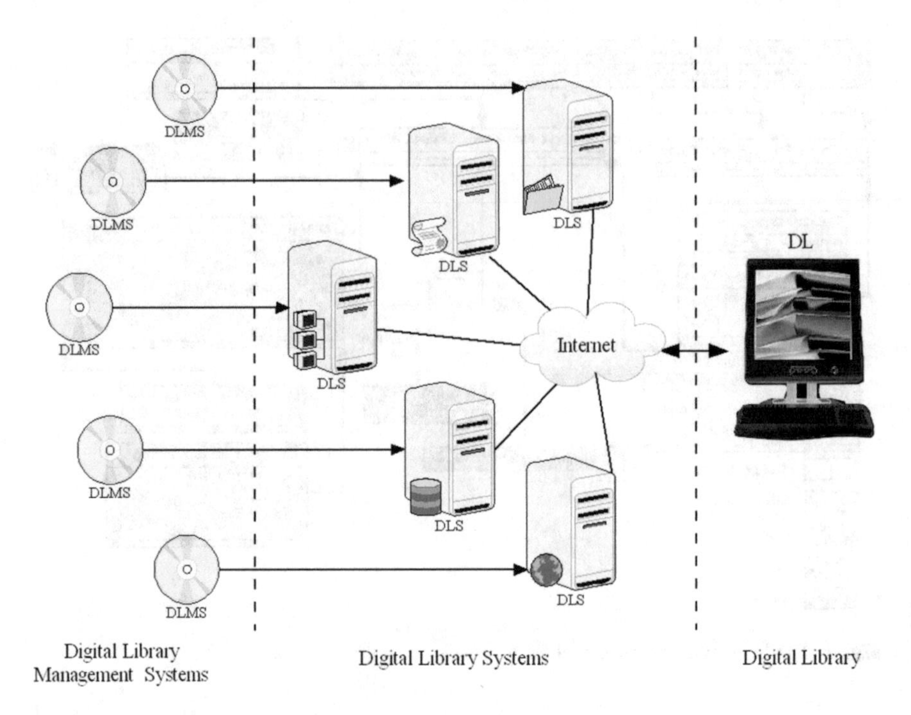

Fig. 7 The three-tier framework composed by DL, DLS and DLMS

the final 'system' actually perceived by the end users, namely the content creators, content consumers and digital librarians.

The relationship between DL, DLS, and DLMS may be therefore expressed analogously to the context of information systems [20]. In detail, a DataBase Management System, or DBMS (e.g. MySQL or PostgreSQL), may be compared to a DLMS as both provide general services of data management. A DBMS along with other supporting distributed software can be compared instead to a DLS. Finally, the union of these elements (DBMS instance + bundled software on one side, DLMS + DLS on the other side) with the data and users constitutes the information system, corresponding to the DL.

The heterogeneity of content and of the resulting functionalities and policies forces the DL designer to adopt a multidisciplinary approach for the concrete realization of the DLs, in order to satisfy the objectives and the needs of different application contexts. In this direction the inclusion of new digital content such as SOs has a twofold implication: (i) on the one hand, it allows DL to acquire a plethora of content creators/consumers which are fundamental building block for the IoT [5]; (ii) on the other hand, it allows the end users to acquire a valuable tool to simplify the complex SOs management, exploiting the functionalities that DLs provide for their traditional content by means of the DLMSs.

As the DL universe evolved, several conceptual models characterizing different DL facets have been proposed and implemented (e.g. the CIDOC Conceptual

Reference Model, the 5S Framework, etc.). In the following subsection, one of the most important DL model, namely the Digital Library Reference Model (DLRM), will be introduced. The next section, therefore, will show how the SOs can be included into DL by matching the features of the proposed SO metadata model and the concepts provided in the DLRM.

5.1 The Digital Library Reference Model

The Digital Library Reference Model (DLRM) [11] is currently the main reference model for architecting DLs. In the following, the DLRM terms are reported in italics. DLRM states that a DL is similar to an *Organization*, which foundations are six core concepts or domains: *Content, User, Functionality, Policy, Quality, and Architecture*. The first five of them capture the features characterizing the DL and its expected services. The Architecture, instead, captures the systemic properties underlying the expected services. The cornerstone of the DLRM as well as the shared concept between the six DL domains is the *Resource*, which models any element easily identifiable through an unique *Resource ID*. As long as the *Resource* complies with the established specifications defined into the *Resource Format* (an arbitrarily complex and structured schema usually drawn from an ontology to guarantee an uniform interpretation), it may be accessed, queried and managed. Figure 8 shows a map of the aforementioned DL concepts.

The *Content Domain* represents the various aspects related to the modeling of information managed in the DL universe to serve the information needs of the active entities interacting within the DL, namely the *Actors*. The main *Resource* of *Content Domain* is the *Information Object*, which is an information item that seamlessly provides data to the DL *Actors* by means of the functionality offered by the DL itself. Such interactions, organized according to different types of criteria, are displayed in different *Views* and recorded by the *Action Log*, so allowing the

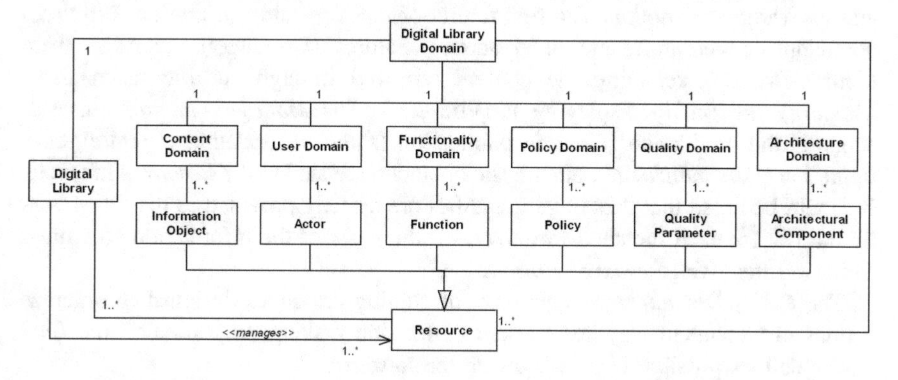

Fig. 8 Digital library main concepts

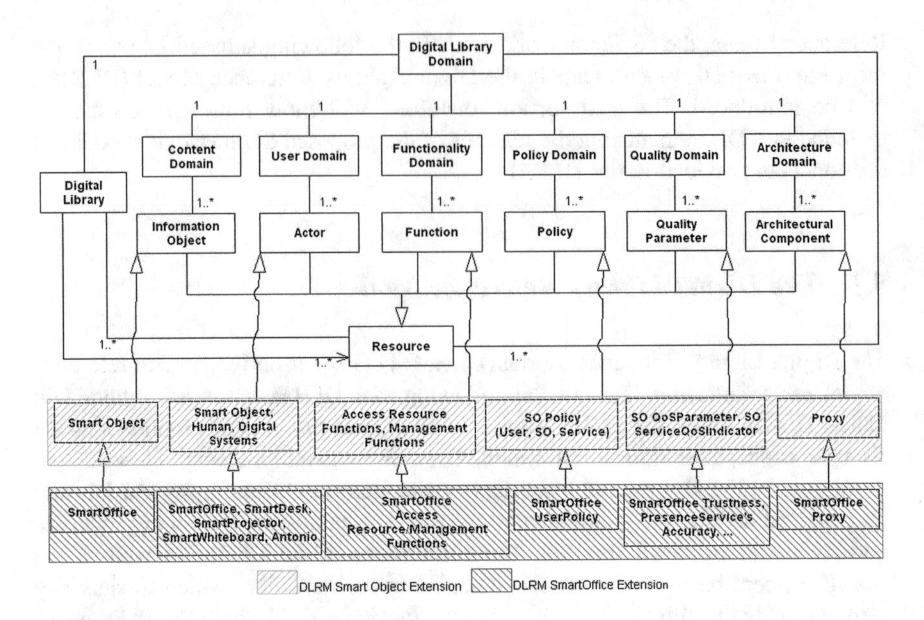

Fig. 9 Digital libraries main concepts mapping

Actor profiling. Finally, *Information Objects* can be grouped into *Collections* concept, a specialization of the *Resource Set*, for some management or application purposes.

The *User Domain* represents the various aspects related to the modeling of entities, either human or machines, interacting with any DL system. In particular, the DL *End-Users* are the ultimate clients the Digital Library is going to serve.

The *Functionality Domain* represents the various aspects related to the modeling of facilities/services provided in the DL universe to serve *Actor* needs. A *Function* is a particular operation that can be realized on a *Resource* upon an *Actor* request. Functions can be specialized in two main classes: the *Access Resource Function* and the *Manage Function*. The first family of functions aims at finding *Resources* compliant to certain (static or dynamic) features (*Discovery*), querying them (*Search-Browse*), retaining the content retrieved through specific mechanisms (*Acquire*) and finally displaying it (*Visualize*). The *Manage Function*, instead, supports the production (*Create*), publication (*Publish*), updating (*Update*), configuration (*Personalize*) and other basic operations related to the *Resource* lifecycle. It should be noted that these functionalities are directly provided to the DL *Actors* by the DL for each included *Resource*, on the basis of the information structured following the given *Resource Format*.

The *Policy Domain* represents a set of guiding principles designed to organize actions in a coherent way and to help in decision making. In particular, the *User Policy* defines possible *User* actions on the *Resource*.

The *Quality Domain* captures the aspects that permit considering DL systems from a quality point of view, with the goal of judging and evaluating them with respect to specific facets. It represents the various aspects related to features and attributes of *Resources* with respect to their degree of excellence. In particular, the DLRM provides *Quality Parameter* on the *Resources* (*Generic Quality Parameter*), on the *Information Object* (*Content Quality Parameter*), and on the *User* (*User Quality Parameter*).

The *Architecture Domain* represents the various aspects related to the software systems that concretely realize the DL universe. In particular, it offers useful insights about how to develop new efficient DL systems and how to improve current ones.

6 Inclusion of Cyberphysical Smart Objects into Digital Libraries According to the Digital Library Reference Model

The inclusion of SOs into DLs is carried out according to the Digital Library Reference Model (DLRM) [11]. As the DLRM is structured into six domains, the SO is contextualized in each of such domains by discussing matching and implications of its inclusion with respect to the adopted SO metadata model, and providing meaningful examples related to the case study proposed in Sect. 4 (Fig. 9).

An SO following the proposed SO metadata model, which is compliant with the definition (and also rationale) provided in the DLRM, can be straightforwardly included as *Resource* in a DL. In fact, it is uniquely identifiable through the *Resource ID* and its *Resource Format* constitutes a representation which makes itself easily accessible, queryable and manageable by the DL entities.

The SmartOffice presented in the case study fulfills such requirements as shown in Table 3 and could be considered an includable *Resource* of the DL, since it has

Table 3 Mapping between the resource DLRM concepts, the Smart Object and the SmartOffice

DLRM concepts	General Smart Object concepts	SmartOffice concepts
Resource	Instance of Smart Object	Instance of SmartOffice
Resource ID	Smart Object FingerPrint	SmartOffice FingerPrint: • Identifier: Office1 • Type: SmartOffice • Creator: SenSysCal • Location: University of Calabria, 41c… • QoSParameter: 0.95 Trustness
Resource Format	Smart Object metamodel	SmartOffice metamodel

an identifying FingerPrint and it complies, together with its aggregate SOs, with the presented SO metadata model. In detail, the SmartOffice FingerPrint includes its creator (SenSysCal), its unique id (Office1), type (Smart Office) and other specific information.

6.1 Content Domain

With respect to the Content Domain, the SO is a novel *Information Object* that contributes to the production and consumption of content that will be handled by the DL *Actors* through the SO Services and the related SO metadata. The latter are well suited to be contextualized in different *Views* and exploited to realize *Action Log* and *Actor profiling*. Moreover, the proposed SO metadata model allows a logical grouping of SOs, in order to build aggregated entity which complements and enhances the Services of the various components SO. The full matching between the *Content Domain* concepts and the Smart Office related ones is shown in Table 4. In fact, as a novel Information Object, the SmartOffice produces and consumes digital contents and aggregates other SOs (SmartWhiteboard, SmartDesk, SmartProjector) which interact with each other and with the SmartOffice itself, without losing their independence or alter their nature. Every service exposed by the SmartOffice presents useful metadata to describe the expected content output (service id, requested parameters, return type, description) and the list of operations implementing the service itself. Moreover, taking the example of the `LightService`, the SmartOffice could maintain the list of the users, who requested such service, and display such information in an aggregate weekly *view* or in a monthly average *view*.

Table 4 Mapping between DLRM content domain concepts, the Smart Object and the SmartOffice

DLRM concepts	General Smart Object concepts	SmartOffice concepts
Information Object	Smart object	SmartOffice
Collection	Aggregated smart objects	Smart Desk, Smart Whiteboard, Smart Projector
Action Log	When and how a specific SO service has been used	"When the LightService has been used?"
Actor profiling	Who used specific a SO service	"Who used the LightService?"

6.2 User Domain

The SOs play a dual role within the DLRM, and specifically within the *User domain*. In fact, SOs are both *content creators*, because they produce or update data and information, and *content consumers*, as it often happens that they are themselves users of other SOs or *Resources* in general. The proposed SO metadata model contemplates both of these occurrences, together with the possibility that other *Actors*, humans or in a broader sense digital systems (such as web server, robot, etc.), can play the role of consumer of SO-generated content. Other *Roles* envisaged in the DL reference model within the *User Domain* (DL Managers and DL Software Developers) are complementary to proposed SO metadata model.

Regarding the case study, as Table 5 shows, the SmartOffice plays both the role of: (i) content creator, in order to provide services like the `LightService` to the human user (e.g. the employee in the Smart Office) or even to general digital systems (e.g. a remote web service or software agent); (ii) content consumer, even exploiting data generated by other SOs like the SmartDesk through the `PresenceService`.

6.3 Functionality Domain

The *Access Resource Functions* and *Manage Functions* provided by the DL may usefully exploit the information structured in the proposed SO metadata model. For example, the *Discovery Function* may use the contents of the fields "creator" and "type" of the FingerPrint category to research all SOs created by a particular subject and belonging to a certain typology (e.g. all the "SmartProjectors" of the "University of Calabria"). Dynamic information about current SO status (e.g. current temperature) can be similarly retrieved by using *Query Function*.

Considering the case study, a possible scenario is described in Table 6, when a *User* of the DL searches for a specific service (e.g. the `LightService`) through the *Discovery Function*, this will query the metadata generated by the *Resources* contained in the DL, among which the SmartOffice. The SmartOffice matches the search criteria, so the User will fulfill a list of parameters contained in the

Table 5 Mapping between DLRM user domain concepts, the Smart Object and the SmartOffice

DLRM concepts	General Smart Object concepts	SmartOffice concepts
Actor	Smart Object, Users (Smart Objects, Humans, Digital Systems)	SmartOffice, Smart Desk, Smart Whiteboard, Smart Projector and Antonio
Content creator	Smart Object	SmartOffice w.r.t. Light Service's user
Content consumer	Smart Object, Users	SmartOffice w.r.t. Presence service's user

Table 6 Mapping between the DLRM functionality domain concepts, the Smart Object and the SmartOffice

DLRM concepts	General smart object concepts	SmartOffice concepts
Access resource functions (discovery-search-query-visualize)	The User exploits the *DL Discovery Function* to discover a specific service; hence, the User submits a request to the Service through the *DL Query Function*	The User queries the DL for a specific services: the DL *Discovery Function* finds that the inserted criteria match with the LightService metadata; so the request is carried out by the *Query Function*
Manage Functions (create-publish-update-personalize)	The User specifies through the *Personalize Function* how to display the SO Services usage	The User specifies through the *Personalize Function* the desired view (daily or monthly) for displaying the LightService request output

Personalize Function (e.g. which *View* to be adopted) before proceeding to the query. Based on such information, the DL will interact with the SmartOffice `LightService` to carry out the requests through the *Visualize Function.*

6.4 Policy

The proposed SO metadata model is neutral with respect to the concept of *Policy*. Few changes to the SO metadata model could be carried out to regulate the interactions between the SO user and the SO services, according to what is present in the reference DL model respectively with the *User Policy* and *Content Policy.* In particular, one could implement the concept of *Policy* by directly associating it to the SO User or SO Service entity, or binding it outside of the SO metadata model, at the level of the DL.

As an example related to the case study, referring to the *User Policy* (see Table 7), it could be stated that the `LightService` could be accessed only by a few trusted users (the ability to close the lights remotely can have

Table 7 Mapping between the DLRM Policy domain concepts, the Smart Object and the SmartOffice

DLRM concepts	General smart object concepts	SmartOffice concepts
User Policy	SO services enabled on the basis of user degree of reliability	SmartOffice w.r.t. to LightService and VisualizationService access

undesirable consequences in case of abuse from malicious users) while the `VisualizationService` could be used by everyone present in the SmartOffice. In this direction, the definition of a user's level of reliability could be a facility for the implementation of the *UserPolicy* (see *UserQualityParameter* in the next subsection).

6.5 Quality Domain

The proposed SO metadata model already contains two elements that refer to the SO quality (QoSParameter) and the quality of the SO Services (QoSIndicator), in full agreement with the DLRM *Generic Quality Parameter* and *Content Quality Parameter.* Regarding the DLRM *User Quality Parameter,* it could be easily imported into the SO metadata model, for example by assigning each SO User reliability value, on the basis of which it is possible to define *Policy* and granting special rights or access privileges to the SO Services.

For instance, as summarized in Table 8, the SmartOffice has a high trustness value (*Generic Quality Parameter*) because it exploits carefully designed and maintained hardware and software components; the `PresenceService`, in turn, has a high accuracy (*Content Quality Parameter*) because the percentage of false positives and false negatives is extremely low.

6.6 Architecture

As opposed to traditional *Resources* that can be acquired and placed directly into DLs (such as documents, videos, etc.), mostly of the SOs stably belong to external systems. The SmartOffice itself falls in such scenario, with its aggregated SOs. Therefore, the inclusion of a SO within the DL architecture presented in the DLRM may involve (i) the coupling of every SO (hence defined *Hosting Node*) with a *Software Component*, suitable designed according to the DLRM, which interfaces

Table 8 Mapping between the DLRM quality domain concepts, the Smart Object and the SmartOffice

DLRM concepts	General Smart Object concepts	SmartOffice concepts
Generic Quality Parameter	SO QoS parameter	Trustness value of QoSParameter
Content Quality Parameter	SO service QoS Indicator	Accuracy value of PresenceService's QoSIndicator

Table 9 Mapping between the DLRM architecture domain concepts, the Smart Object and the SmartOffice

DLRM Concepts	General Smart Object concepts	SmartOffice concepts
Inclusion strategy	Proxy-based inclusion	Proxy-based inclusion

the specific SO features and SO services with the DL requirements and functionalities, or (ii) the creation of a new component, currently not present in the reference architecture, delegated to the on-demand SOs virtualization [21, 22]. In order to make SOs look "transparently" integrated into the DL like other traditional *Resources,* but at same time to design a loosely coupled integration (maintaining the SO design neutral with respect to the application scenario) the most viable solution is to exploit a proxy within the DL architecture that virtualizes the SOs before their inclusion (Table 9).

7 Conclusion and Future Work

In this paper we have proposed an approach for the inclusion of SOs into DLs compliant with the DLRM defined by DL.org [12]. The inclusion is based on a metadata model for SOs purposely defined to fully characterize all SO properties (both physical and cyber) as well as their interactions with other human, digital and cyberphysical actors. The approach has also been exemplified through a case study concerning a smart office environment. In particular, the SO metadata model has been instantiated with respect to the case study, and the resulting SmartOffice model has been used to exemplify the main SO inclusion concepts.

Such an inclusion would enable, from one perspective, to effectively support discovery, querying and management of SOs through tools and facilities provided by modern DLs and, from another perspective, to extend currently available DLs with a new type of object to collect, manage and preserve. To the best of our knowledge, our proposal is the first research aiming at this inclusion that would pave the way towards the development of cyberphysical DLs.

Future work directions will be mainly twofold: (i) addressing interoperability and trust issues of cyberphysical DLs [23]; (ii) implementing the proposed approach in a real DL management system such as Fedora [24] and/or DSpace [25].

Acknowledgement This work has been partially supported by DICET INMOTO Organization of Cultural Heritage for Smart Tourism and REal Time Accessibility (OR.C.HE.S.T.R.A.) project funded by the Italian Government (PON04a2 D).

References

1. Ross, S.: Digital library development review. Final report. National Library of New Zealand (2003)
2. Chowdhury, G., Sudatta, C.: Introduction to Digital Libraries. Facet Publishing, UK (2002)
3. Saracevic, T.: Digital library evaluation: toward evolution of concepts. Libr. Trends **49**(2), 350–369 (2000)
4. Amato, G., Gennaro, C., Rabitti, F., Savino, P.: Milos: A multimedia content management system for digital library applications. In: Research and Advanced Technology for Digital Libraries, pp. 14–25. Springer, Berlin (2004)
5. Kortuem, G., Kawsar, F., Fitton, D., Sundramoorthy, V.: Smart objects as building blocks for the internet of things. IEEE Internet Comput. **14**(1), 44–51 (2010)
6. Atzori, L., Iera, A., Morabito, G.: The internet of things: A survey. Comput. Netw. **54**(15), 2787–2805 (2010)
7. Fortino, G., Guerrieri, A., Russo, W., Savaglio C.: Middlewares for Smart Objects and Smart Environments: overview and comparison. In: Internet of Things Based on Smart Objects: technology, middleware and applications. Internet of Things: Technology, Communications and Computing, pp. 1–27. Springer, Berlin (2014)
8. Fortino, G., Guerrieri, A., Russo, W.: Agent-oriented smart objects development. In: IoT and Logistics Workshop jointly held with 16th IEEE International Conference on Computer Supported Cooperative Work in Design (CSCWD 2012), pp. 907–912. IEEE, Wuhan (China) (2012)
9. Wu, F.J., Kao, Y.F., Tseng, Y.C.: From wireless sensor networks towards cyber physical systems. Pervasive Mobile Comput. **7**(4), 397–413 (2011)
10. Fortino, G., Rovella, A., Russo, W., Savaglio, C.: Including Cyberphysical Smart Objects into Digital Libraries. In: Internet and Distributed Computing Systems, pp. 147–158. Springer International Publishing (2014)
11. Candela, L., et al.: The Digital Library Reference Model. Report. http://www.dlorg.eu/index.php/outcomes/reference-modeloutcomes/reference-model (2010)
12. Dl.org—Digital Library Interoperability, Best Practices and Modelling Foundations. http://www.dlorg.eu/
13. Uckelmann, D., Harrison, M., Michahelles, F. (eds.): Architecting the Internet of Things. Springer, Berlin (2011)
14. Kawsar, F., Nakajima, T., Park, J.H., Yeo, S.S.: Design and implementation of a framework for building distributed smart object systems. J. Supercomput. **54**(1), 4–28 (2010)
15. Serbanati, A., Medaglia, C.M., Ceipidor, U.B.: Building blocks of the internet of things: state of the art and beyond. In: Turcu, C. (ed.) Deploying RFID-Challenges, Solutions, and Open Issues. InTech (2011)
16. Fortino, G., Lackovic, M., Russo, W., Trunfio, P.: A discovery service for smart objects over an agent-based middleware. In: Pathan, M., Wei, G., Fortino, G. (eds.) Lecture notes in computer science, LNCS, vol. 8223, pp. 281–293. Springer, Heidelberg (2013)
17. Fortino, G., Russo, W., Rovella, A., Savaglio, C.: On the classification of cyberphysical smart objects in internet of things. In: International Workshop on Networks of Cooperating Objects for Smart Cities 2014 (UBICITEC 2014), vol. 1156, pp. 76–84 (2014)
18. Fortino, G., Guerrieri, A., Lacopo, M., Lucia, M., Russo, W.: An agent-based middleware for cooperating smart objects. In: Corchado, J. (ed.) Highlights on Practical Applications of Agents and Multi-Agent Systems. Communications in Computer and Information Science (CCIS), vol. 365, pp. 387–398. Springer, Heidelberg (2013)
19. Suleman, H., Edward, A.: Designing protocols in support of digital library componentization. Research and Advanced Technology for Digital Libraries, pp. 568–582. Springer, Berlin (2002)
20. Connolly T.M., Begg, C.E.: Database Systems: A Practical Approach to Design, Implementation, and Management. Pearson Education, New York (2005)

21. Fortino, G., Guerrieri, A., Russo, W., Savaglio, C.: Integration of Agent-based and Cloud computing for the Smart Objects-oriented IoT. In: IEEE Computer Supported Cooperative Work in Design (CSCWD), Taiwan (2014)
22. Kelaidonis D., et al.: Virtualization and cognitive management of real world objects in the internet of things. In: IEEE International Conference on Green Computing and Communications (GreenCom). IEEE (2012)
23. Innocenti, P., Vullo, G., Ross, S.: Towards a digital library policy and quality interoperability framework: the DL.org project. New Rev. Inf. Netw. **15**(1), 29–53 (2010)
24. Fedora Project. http://www.fedora-commons.org/about
25. DSpace. http://www.dspace.org

Cooperation of Smart Objects and Urban Operators for Smart City Applications

Simona Citrigno, Sabrina Graziano and Domenico Saccà

Abstract The project "TETRis—TETRA Innovative Open Source Services" has delivered a technological infrastructure for enabling innovative services for Smart City/Smart Territory. This paper describes the software tools and intelligent platforms for collecting, representing, managing and exploiting data and information gathered from sensors and devices deployed in the territory. Tools and platforms are integrated into a complex smart environment that provides advanced services to citizen and operators for environmental monitoring, urban mobility and emergency management. Although the project is mainly based on the utilization of the communication protocol TETRA, the application scenarios may work with other network protocols as well.

Keywords Smart objects · Urban monitoring · Urban mobility · Intelligent platforms · Wireless sensor networks

1 Introduction

The concept of smart city is used all over the world with different nomenclatures, context and meanings [1–3]. The concept adopted in this paper is: a city is smart if it acts in a forward-looking way in economy, people, governance, mobility, environment, and living, using a suitable combination of endowments and activities of

S. Citrigno (✉)
Centro di Competenza ICT-SUD, Rende (CS), Italy
e-mail: simona.citrigno@cc-ict-sud.it

S. Graziano
OKT Srl, Rende (CS), Italy
e-mail: sabrina.graziano@okt-srl.com

D. Saccà
Università della Calabria and ICT-SUD, Rende (CS), Italy
e-mail: sacca@unical.it

A. Guerrieri et al. (eds.), *Management of Cyber Physical Objects in the Future Internet of Things*, Internet of Things,
DOI 10.1007/978-3-319-26869-9_8

self-decisive, independent and aware citizens. Therefore, a smart city monitors and integrates conditions of all of its critical infrastructures, optimizes its resources, plans its preventive maintenance activities, and monitors security aspects while maximizing services to its citizens [4]. To be smart, a city must interconnect the physical infrastructure, the ICT infrastructure, the social infrastructure, and the business infrastructure to leverage the collective intelligence of the city [5, 6]. The use of advanced ICT technologies is crucial to make its infrastructure components and services (including city administration, education, healthcare, public safety, real estate, transportation, and utilities) more intelligent, interconnected, and efficient and to identify new, innovative solutions to city management complexity, in order to improve sustainability and livability [7]. In sum, a smart city must strive to make itself "smarter" (more efficient, sustainable, equitable, and livable) [8].

The activities described in this paper are related to the design and prototypal implementation of innovative services for an intelligent management of an urban territory within novel smart city application scenarios. A number of solutions and advanced technology platforms have been identified for enabling the various entities operating in the area of interest (municipalities, provinces, regions, universities, etc.), as well as citizens and urban operators, to effectively cooperate for an efficient usage of urban resources. In fact, the wealth of information acquired on the territory through the usage of sensors and devices interconnected by local and remote communication systems are suitably stored and processed to support high added-value services to improve the quality of the territory itself in terms of livability and sustainability. A crucial aspect is the involvement of citizens and urban operators who become the main tutors of the territory, the so-called "social sensors", for the detection of critical situations in the urban territory.

The innovation scenarios and solutions described in this paper have been realized within the project PON 2007–2013—Research and Competitiveness "TETRis—TETRA Innovative Open Source Services" according to reference general frame of "Internet of Things" for supporting Smart City/Smart Territory [9], in which the acquisition of data by objects is applied to large territorial areas by exploiting the widespread availability of communication networks [10, 11]. The collected data, properly enhanced and enriched, foster innovative services oriented to the production and exchange of knowledge among the different actors interconnected in urban and regional networks. The development of these services has been realized through the cooperation of smart devices and objects as well as of operators and users of the services themselves.

A key role in Internet of Things, as well as in smart city scenarios and services, is played by the concept of *smart object*, first introduced in [12], which is a physical/digital object having a unique identifier that is used to digitally manage physical things (e.g., sensors), to track them throughout their lifespan and to annotate them (e.g., with descriptions, opinions, instructions, warranties, tutorials, photographs, connections to other objects, and any other kind of contextual information imaginable), and to consciously handle its relationships with other smart objects and with remote systems. In sum, a smart object is a physical/digital object augmented with sensing/actuating, processing, and networking capabilities that may embed human behavioral logic [13].

Smart objects are typically part of a *Smart Environment*, which is "a physical world that is richly and invisibly interwoven with sensors, actuators, displays, and computational elements, embedded seamlessly in the everyday objects of our lives, and connected through a continuous network" [14]. Smart Environments are often based on a suitable middleware that enables communication and management of smart objects in distributed applications [15–17].

The activities described in this paper refer to two main application scenarios identified within a Smart City context: (A) Urban Mobility (B) Territory Monitoring, Control and Maintenance [18].

The remainder of the paper is organized as follows: Sect. 2 presents an overview of the TETRis project and describes its main goal, Sect. 3 illustrates two meaningful innovative application scenarios for smart city/territory, Sects. 4 and 5 focus on the scenario respectively of urban mobility and of urban monitoring and risk analysis and Sect. 6 withdraws the conclusion and discusses further work.

2 Project Objectives of TETRis

The TETRis project main goal is to create high value-added services by exploiting and possibly extending the functionalities of the Terrestrial Trunked Radio (TETRA) communication system. The TETRA is an open standard for mobile radio communications developed by the European Telecommunications Standards Institute (ETSI) and specifically designed to support Professional Mobile Radio communications (PMRs) in a number of market segments such as public safety, transportation, utilities government, military, commerce, industry. TETRA is deployed in over 88 countries worldwide, and the main market is by far that of national public safety organizations. The primary goal of public safety is to carry out all the necessary actions for the prevention and protection from events, such as dangers, injuries, or damages, that could threaten the safety of the general public.

TETRA system is particularly suitable to meet the needs of professional users of emergency care organizations dealing with public utilities, public security forces, transport companies, and it represents also the answer to solve the growing needs of private mobile radio systems (PMR) users, both in terms of radio traffic decongestion and in enhancing voice and data services. TETRA is designed to provide operational and service communications between land mobile units, naval and air and their related control centers, in either voice and data modes, and it integrates with existing radio communication frameworks and commercial communication systems, thus ensuring maximum efficiency both in terms of transmission resources and in management and use.

TETRA provides a common and standard infrastructure for secure and reliable communications and it also supports unique features such as group conversations, dispatcher centers, and direct communications. While the initial focus of TETRA has been on voice communications, data communications have been supported since the beginning and nowadays are gaining more importance. The prevention and management of critical situations related to public safety, requires the real-time acquisition of data from the field in order to react more consciously, faster and better.

The project acts along three main axes in order to enhance TETRA functionalities and to extend its usage to novel application domains:

1. The evolution and the opening of novel application fields for TETRA in order to define new information services for operators, exploiting new models and open source tools for the interconnection of TETRA with other networks and the identification of new type of TETRA-based devices suitably interoperated with existing sensors and sensor networks;
2. The modeling and prototyping of an Open Source framework implementing a Smart Objects cooperation model and enhancing their management by means of so-called Smart Environments;
3. The identification of novel scenarios and application models in the perspective of Smart City/Smart territory services applied to territory monitoring, emergency management and intelligent support to urban mobility.

The specific objectives of TETRis project can be read as follows:

- Bringing economic and social benefits to the community through more targeted and effective actions by Public Administration and Public Security operators in application scenarios such as emergency management, environmental protection, mobility and services to citizens, with the contribution of the citizens themselves through the sharing of information and the use of innovative tools for social networking;
- Extending the pervasiveness and effectiveness of public administration services, instrumental bodies, local police, health operators, transport companies in the reference areas; Improving the quality of life and the sense of safety of citizens through the spreading of safe and reliable technology infrastructure "always on".

A sketch of the TETRis architecture is shown in Fig. 1, that includes the following elements: networks (e.g., sensors networks, TETRA networks, commercial communication networks), providing communication among Smart Objects for the collection of data from the environment and for the management of some intelligent devices; Smart Objects, that coordinate and manage the interaction among Smart

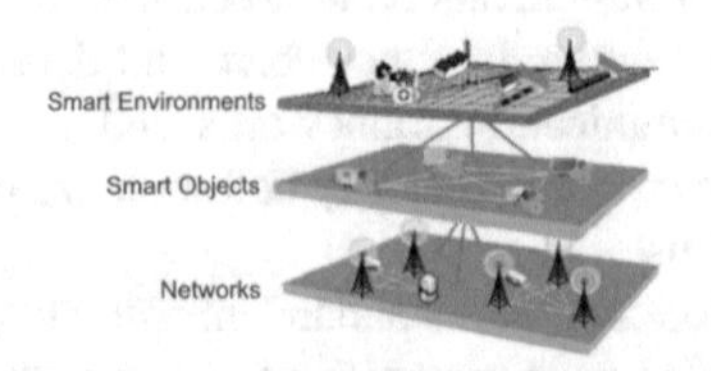

Fig. 1 TETRis overall scenario

Objects themselves and provide added value services to the upper level; Smart Environments, that produce knowledge through the continuous and collaborative interaction of the different elements belonging to the other levels.

The project activities are organized in a number of work packages, each of them consisting of a number of Industrial Research and Experimental Development activities.

During the experimental phase, a TETRA test bed infrastructure has been setup for testing data transmission over a TETRA network. A number of tests have been carried out using a TETRA terminal, a Wi-fi network, some sensors and devices. Both sensors and devices interacted with a gateway node equipped with specific interface for the transfer of the acquired data.

This paper describes the experimental activities conducted by the project in the following two application areas:

- TETRis Smart Environments for mobility—it concerns the implementation of a model for the detection of mobility problems in urban areas through the usage of specific Smart Objects located in the territory and the usage of a network of sensors connected to them. Data are collected and aggregated into a data warehouse feeding a Mobility Intelligence platform defined through the design of innovative techniques of space-temporal data analysis and mining of complex data, including trajectories.
- TETRis Smart Environments for territory monitoring and delivery of services to citizens—it defines a Smart Environment managing a network of physical sensors connected to Smart Objects that is enriched by “social” sensors which detect in real time the status of the territory. These so-collected data are stored and aggregated into a data warehouse feeding a Territory Intelligence platform, which enables the extraction and processing of knowledge for monitoring the territory.

Within the project two collaboration agreements have been subscribed with the municipalities of two towns in Southern Italy: (i) Cosenza as for the Urban Mobility area, and (ii) Rende as for the Urban Monitoring area. The two municipalities have shown a high interest in the experimentation of innovative IT solutions and techniques in their view of pursuing the realization of a new model of a city. A smart city seen as an intelligent system that supervises on the compliance and control of the environment and effectively manages resources through the use of technological ICT infrastructure and innovative tools in order to deliver high added-value services for citizens.

3 Design of the Applications Scenarios for Smart City

The two scenarios that have been identified for a typical application in support of a Smart City, Urban Mobility and Territory Monitoring, Control and Maintenance, share the same basic architecture that includes the following elements:

- *Actors*—divided into three categories:
  - *Government Body*, which is responsible for managing the Smart Environment and the Smart Object network distributed throughout the area;
  - *External Local Authorities*, which are interested in the services delivered by the Smart Environment and Smart Objects with which they interact;
  - *Individuals*, who can be either citizens or workers (operators) performing their duties in the area of interest under the direction of the managing body;
- *Government central systems*, which are responsible for the overall application functioning and for the control of the Smart Objects network distributed on the territory;
- *External Local Authorities systems*, which interact with the Government central systems and the Smart Objects on the basis of operating protocols agreed with the central systems and that can also interact with other external authorities systems;
- *Smart phones APPs*, used by individuals;
- *Smart Objects*, which are distributed on a territory under the management of the central system, and perform two types of communication:
  - *Remote communication*: (i) with the Government central system through TETRA, (ii) with External Local Authorities systems through commercial telecommunications protocols (GSM, UMTS, etc.) and possibly also with TETRA;
  - *Local communication*: (i) with operators through WiFi and NFC technology for instant activation and personalization of the interaction, (ii) with citizens through WiFi, (iii) with physical objects, related to a Smart Object and distributed over the territory, typically using ZigBee protocol or RFID.
- *Sensors*, which can be:
  - classical traffic detection sensors, related to a Smart Object, connected together with ZigBee network protocol or RFID;
  - classical sensors for environmental conditions detection (temperature, humidity, sound, CO, dust, etc.) connected together with an ad hoc network;
  - social sensors, that means citizens and operators having a mobile device provided with applications for signaling and/or monitoring events that occur in the territory of interest;
  - devices, sensors/actuators of urbotic (i.e., automation at urban level) that can either directly interact with Smart Objects using low power protocols but with high performance, such as Cliffside and Wibree, or through a dedicated control kit using WiFi connection.

Figure 2 shows the overall picture of a typical scenario of Smart Objects usage within a Smart City with an indication of the various components and their interactions. Smart Objects can make use of remote communication protocols, such as GSM, TETRA and/or Internet, for information exchange with Government Body and External Local Authorities concerning urban monitoring and emergencies management. Smart Objects may also exchange data with citizen and operators

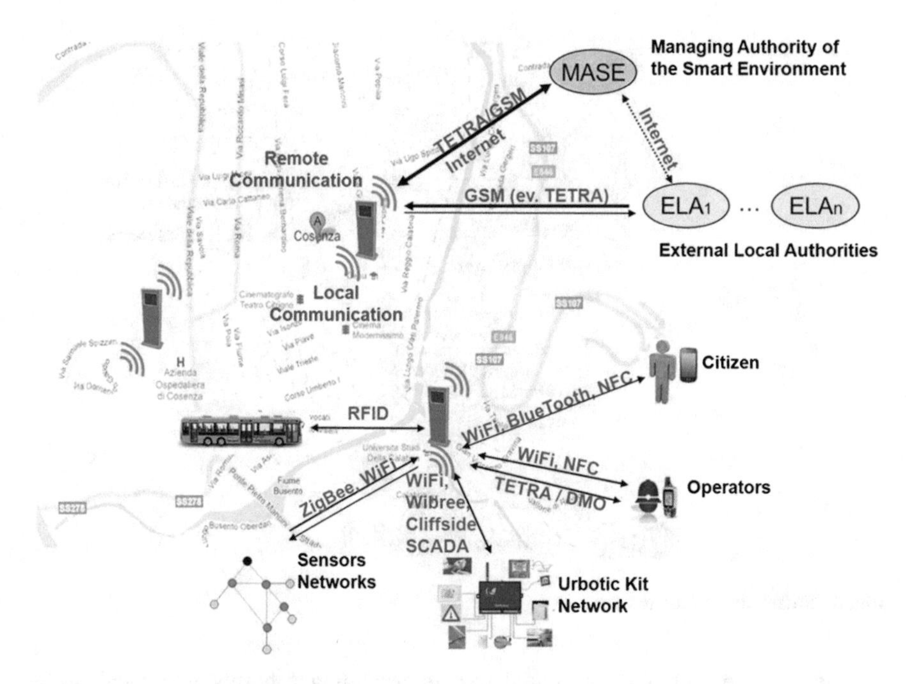

Fig. 2 A smart object usage scenario

making use of local communication protocols, such as: WiFi, Bluetooth and NFC, when interfacing with citizen smartphone applications; WiFi, NFC, TETRA, when communicating with operators devices; RFID, for bus mobility detection; WIFi, ZigBee, Cliffside for sharing information with sensors networks.

Figure 3 shows a general architecture of a Smart Object, in which the following relevant component is identified: the Smart Object software intelligence, which can collect data coming from sensors networks, citizens and operators apps, then it processes such data on the basis of some alerting and control criteria, and in the end it delivers the results to some external entities using local and/or remote communication protocols and/or specific interfaces, depending on the target users. The Smart Object architecture can vary depending on the scenario taken into consideration and it can be seen as a three level architecture with the following tiers: (1) a basic level where social sensors can communicate information through the usage of smartphone apps, and physical sensors able to detect some specific determined measures, and continuously transmitting data to the government central system; (2) a middleware level, which is able to collect data transmitted by sensors and which is provided with an alerting system. Data collected at this stage feed operational and informational dashboards that are used by citizens and operators; (3) a business intelligence system, which integrates data coming from different internal and external sources and provides territory, mobility and security intelligence services to Government Bodies and External Local Authorities by means of decision support dashboards.

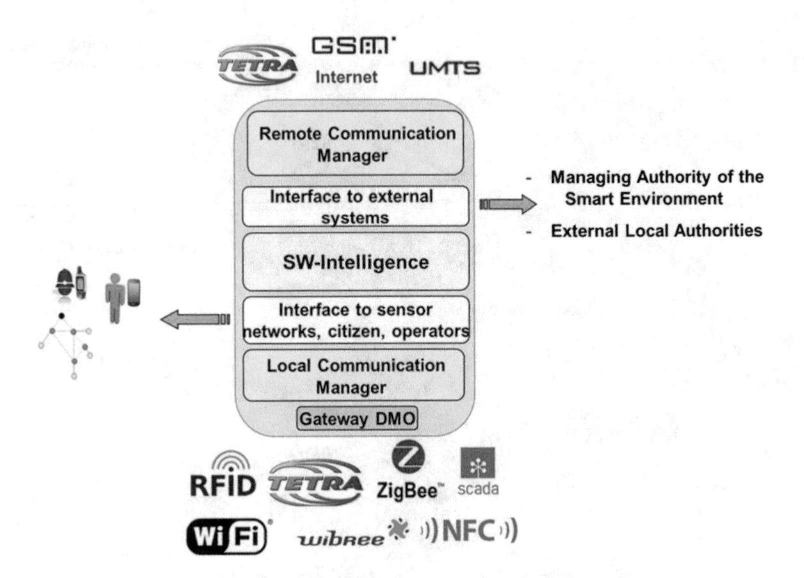

Fig. 3 Smart object architecture

The smart object plays a crucial role in providing a highly integrated urban network to overcome the drawbacks of current fragmentation and heterogeneity of telecommunication technologies and reduced support for ubiquitous connectivity and coverage. The problem of supporting a seamless integration of multiple services/data from multiple applications or on the social cooperation of multiple users has been in more general terms addressed in [19], which proposes an "evolutionary" solution for deployment, extension and management of the network infrastructures in a smart city. This solution handles the heterogeneity of devices and network technologies and the fragmentation of coverage and connectivity in urban areas by means of a so-called "STEM-Net", that consists of dynamically configurable nodes with self-management capabilities. Similar with a stem cell, a stem node can undergo mutation in order to fulfill a given new task, using both built-in node's capabilities and additional capabilities that are dynamically learnt by cooperating with other nodes. The solution is interesting and very promising even though a number of "hard" software challenges need to be taken into account and in-depth further dealt with: in particular how to define and deploy the initial intelligence into the stem nodes and how to equip them with effective learning capabilities. Our smart object can be thought of as a very specific implementation of some of the features of a stem node: most of its intelligence and "mutation" capabilities are externally controlled by the smart environments, playing the role of experts in specific topics. Our approach is less ambitious but the history of Artificial Intelligence (AI) has experienced that the development of new and successful technologies first arises in those cases where a specific knowledge domain is handled.

4 Urban Mobility

The Urban Mobility scenario implements a model based on a network of smart objects connected with some sensors networks, through which it is possible to provide data in real time related to the mobility in an urban area, which can address and support Mobility managers in solving traffic and mobility management issues. The model also includes amenities to deliver specific services to operators and citizens through the usage of mobile devices.

Data from sensors and smart objects, geographically distributed in the urban area, are collected and aggregated into a data warehouse which feeds a Mobility Intelligence platform that performs on-line knowledge-based analysis for the delivery of real-time information to both urban operators and citizens about the smart management and utilization of mobility systems.

In particular, the Mobility Intelligence platform processes data in real time, according to the most recent lines of research in OLAP systems analysis and data mining, and performs as a decision support system supporting operators to extract, quickly and in a flexible way, all the information needed to meet citizens mobility needs [20]. A dashboard offers the possibility to the interested authorities to take strategic/operational decisions and to plan interventions on road infrastructures, public transport routes, parking spaces and multimodal communication infrastructures with citizens. The analysis of data through the intelligence platform can also offer a valid support for the definition of the urban traffic plan through a calibration activity of the model so that it can be possible to update the plan itself and to adapt it to the observed actual reality.

Thanks to an effective integration of data mining tools and geo-referenced data analysis, the platform is also able to depict spatial data, models and results on a geographical map.

The experimental phase has been carried out on the territory of the municipality of Cosenza along three main macro areas of development:

- Urban traffic detection and management, by means of stationary smart objects placed at critical points in the city of Cosenza, which collect frames and videos from webcams and from traffic detection sensors. The smart object used for this experimental phase is made up of a video camera and a computing software unit through which it is possible to detect in real time traffic conditions and to have a traffic estimation thanks to the video frames analysis;
- Detection and management of any violation made by cars on bus corridors, by means of moving smart objects installed on buses of the public city transportation company. The smart object used for this experimental phase is made up of a webcam, a GPS receiver and a computing software unit. Through this system it is possible to monitor the presence of cars on dedicated bus routes and to recognize, through an OCR (optical character recognition) module, their number plates and the geographical coordinates of car location. All data can be made available to traffic operators for penalties and fines application;

- Analysis of the mobility in the city of Cosenza for identifying specific key aspects related to mobility needs and transportation offers in order to implement a set of services for extracting new knowledge that can be useful for Public Transport Managers.

Moreover, within the experimental phase, the following classes of applications for citizens and operators have been also implemented:

1. APP for citizens, providing information on
   - bus tracking: bus arrival times, bus timetable, the position of the nearest bus stops. The information is available on a map and citizens can also create a route on the map depending on the starting and arrival points and on the available bus lines on that route; This specific app named InfoBus has been published and is available on Apple App Store;
   - available parking places: their location on the map, the number of vacancies and the way to reach them based on the geographical position of the citizen requesting such information;
   - real-time traffic updates and traffic jam reports; the traffic condition is highlighted by different colours on the map depending on the level of congestion; hazards and police activities are also proposed;
   - the city life: showing places of interest, restaurants, hotels and clubs.

   All information is contextualized according to the geo-localization of citizens.
2. APP for operators able to:
   - enabling them to monitor traffic and public transport status, to analyze alerts about traffic jams, parking in double row, and other disruptions; Operators can send information about a specific event, its location (latitude, longitude), date and time, adding also a picture about the event and some related notes;
   - issuing fines related to the violation of the traffic code made by cars; Operators can record information about the geo-localization, date and time, the rule that has been violated and they can also attach a picture as a proof of the violation.

The same services delivered by the apps are also available to citizens and operators through a specific Web portal.

Regarding the analysis of the mobility in the city of Cosenza, to define the correct methodologies and processes for urban mobility, the M-Atlas framework has been used [21–23]. M-Atlas is a mobility querying and data mining system centered onto the concept of trajectory. Besides the mechanisms for storing and querying trajectory data, M-Atlas can also perform mechanism for mining trajectory patterns and models that, in turn, can be stored and queried. The knowledge discovery process based on these kind of data help users to answer to specific questions about mobility: e.g. what are the frequent patterns of people's travels? How big attractors and extraordinary events influence mobility? How is it possible to predict intense traffic areas in the near future? How is it possible to characterize traffic jams and congestions?

M-Atlas is equipped with a querying and mining language and some mechanisms to master the complexity of transforming raw GPS tracks into mobility knowledge. Moreover, M-Atlas include (i) the privacy-preserving data publishing and mining techniques, designed to transform trajectory datasets into anonymous forms so that strong privacy-protection is guaranteed together with high data utility; (ii) the analysis of different forms of mobility data, such as mobile phone call records, characterized by complementary weaknesses and strengths in relation to GPS trajectories.

Based on M-Atlas features, the following further analysis have been conducted:

1. An assessment to check the actual possibility to replace part of private mobility in the urban area of Cosenza by public transport together with highlighting deficiencies, waste of resources and suggesting improvements/upgrades. The objective was to verify the capability of Public Transport to satisfy user mobility needs. For this analysis, the GPS logs of buses and GPS tracks of private vehicles were used. Such data have been respectively provided by Amaco S.p.a., the Cosenza Public Transport Company, and by Octotelematics S.p.a.
2. A survey about public bus frequent moving patterns in their routes by checking the mining of bus logs and trajectories. This is to verify time deviations between real travel times and official time tables in order to highlight chronic delays in the service.
3. A reachability evaluation of the city and surrounding areas, using data on both public transport routes and bus logs to compute the actual time distance among the various areas of Cosenza during the day. The objective was to understand how much different areas of the city are served by Public Transport considering different times of a day.
4. A profiling of the population mobility using GSM data by Wind Telecom S.p.a., a mobile phone company, in a specific period of time, to detect how phone callers move among the various city areas and to classify them on the basis of their behaviors. The aim was to identify important categories of people estimating their segmentation in order to evaluate the corresponding services demand.

The results of the analysis have been tested on the field, allowing a continuously monitoring of the general status and health conditions of urban traffic, in terms of impact of Public Transport, actual mobility demand, and mobility profiles of citizens living in the area.

In particular, regarding the first analysis the Public Transport system was mapped to a spatio-temporal network, where nodes were bus stops labeled with name and position and edges were the connections labeled with origin-destination stops and timestamp. Then, the GPS tracks on the Public Transport network was also mapped and it was computed the shortest way to satisfy users' mobility using an agent-based algorithm simulating human mobility in a network [22].

It was found that the delay distribution was affected by the seasonality: in summer the average delay was about 29 min (with a variance of 26), while in winter was about 16 min (with a variance of 15). Going back to the trajectory data and extracting the starting points of users which are not served by the public transport, it was possible to discover which areas were disconnected from the network. By using a clustering algorithm on the starting points of GPS tracks that were not fully covered by the Public Transport, two peripheral areas not reached by the bus service were identified, one industrial and one residential. This result suggested the possible introduction of new lines or the addition of new bus stops to an existing line passing by those peripheral areas. This analysis was very effective in discovering the real needs of the population and how the network can handle them, and it was also useful to highlight potential customers which can be served by the public transport and therefore good candidates for a specific marketing campaign.

As a result of the second analysis, it was found that particular areas of the city could be reached by the Public Transport in a fixed amount of time only in specific time slots. Based on this result, the Public Transport manager could think about adding further bus lines or modifying existing bus schedule and analyzing the impact of this choice in the Public Transport system.

The third analysis showed that almost all buses were late in a range [10, ≠10] min. This kind of information is very useful to spot problems in the buses management, i.e. to improve the service or to highlight too strict schedules which can't be respected by buses in reality.

In the end, the analysis of users calling behavior was useful to classify them into three categories: *Resident*, *Commuters* and *Visitors*. People that appeared only once (i.e. that made only one call in the whole period of observation) before disappearing, were separately classified in the *In Transit* category. The following segmentation was obtained: 23.12 % *Residents*; 14.56 % *Commuters*; 26.45 % *Visitors*, and 28.74 % *In transit*. The 7.13 % were unclassified due to their unclear profile.

This work and the collaboration with the public administration, led towards the definition of a sort of dashboard for a mobility manager made up of a set of end-user services and indexes to evaluate the transport system of a city.

5 Urban Monitoring and Risk Analysis

The Urban Monitoring and Risk Analysis application scenario activities concerns the definition of a Smart Environment integrated with a network of smart objects located throughout the territory for the management and maintenance of the urban environment and for the delivery of timely and innovative services to citizens. During the experimentation phase, a number of physical and "social" sensors have been deployed in the territory to detect in real time the status of the territory itself and the associated risks. The scenario includes a Territory Intelligence platform that, first collects data supplied by smart objects and social sensors and, then, as a result of the appropriate phases of integration and data processing using advanced techniques of data warehousing, it makes the information available to various parties with which the platform interacts in order to properly monitor urban areas.

Within this scenario a number of contextual applications have been made available to citizens and operators working for territory maintenance through their mobile devices. These applications allow them to immediately communicate with the public administrations and utilities in order to report critical situations, malfunctioning in urban networks, road failures and intervention requests for restoring the conditions of good livability of the territory. Events to be reported are classified into the following categories: dangers on the road, stray animals, services and networks, urban refurnish and green, waste depot. The collected data and reports can be also inquired through a web portal, which can be seen as a managing console for performing the following activities: assigning the requests to an operator or a group of operators for the appropriate interventions; monitoring the status of the actions taken; checking the status of the smart objects distributed in the area of interest; implementing an overall urban safety system by an immediate involvement of urban operators.

The experiments were carried out on the territory of the Municipality of Rende focusing on the monitoring of urban and extra-urban areas through real-time information coming from sensor networks connected to smart objects suitably distributed in the urban territory.

Real-time information from smart objects and social sensors feed the model used to draw up new plans of action, or to update/modify the current ones so that a local government can better relate resource planning to the city needs in a dynamic and timely way. Sensors distributed throughout the territory have been also designed to detect the exceeding of prefixed "thresholds", which require contingent changes to maintenance plans by the government bodies.

Within the Urban Monitoring scenario, a Risk Analysis model has been also implemented for road traffic monitoring in urban areas, for the identification and assessment of risks hanging over the system under observation and for taking the necessary countermeasures. The identification of the possible risk events and the activities for monitoring them can enable, in a more easy and rational way, local authorities to implement action plans to prevent the occurrence of risks and/or to reduce their impact, to promptly detect ongoing risks and to alert urban operators for their immediate intervention. The involvement of citizens both in the detection of critical situations and in the immediate assistance has been also taken into account.

The experiments within Urban Monitoring scenario have been conducted also through the usage of wireless sensor networks for monitoring the surrounding environment [24] designed in a way that every sensor is able to forward the acquired information to a collection center (a sort of a gateway smart object), a kind of a multi-hop ad hoc network. This converge-cast communication schema is particularly suitable for collating data on a territory and for communicating them to a central sink (a smart object), in order to make data available for other further data processing. The adopted protocol is CTP—Collection Tree Protocol [25]. Every sensor periodically communicates data to its parent (selected on the basis of the communication protocol policy) going up to the sink in order to forward data to the Territory Intelligence platform.

Two main experimental domains have been identified:

1. Indoor working environment for air quality monitoring;
2. Structural buildings monitoring.

In both domains two wireless sensors networks have been installed within the same building for the experimental activities. In particular, both wireless sensors networks were made up by using two of the most common typology of sensor nodes: TeloSB and Magonode—MNA Board (the last one was used for prototyping and debugging purposes). Such networks were running for 24 h in different days and were made up of 20 nodes in total powered by 2 × 1.5 V AA alkaline batteries. As communication protocol stack it was used the Collection Tree Protocol (routing layer) on top of Low Power Listening (MAC layer). Both protocols were implemented using the TinyOS 2.1.2 open source software. TinyOS is an operative system dedicated to wireless sensors networks developed by Berkeley University. The open-source character of the software decreed its success: the possibility to modify, reuse and adapt the source code to fulfill the developers needs made TinyOS very popular and rich of applications and drivers for a great variety of hardware modules.

Sampling per node occurred every 15 min, a sufficient range of time to provide meaningful measurements. Moreover, all protocols have been tuned for the optimization of the energy consumption, and consequently the network lifetime, in view of the sampling requirement.

In the first domain, air quality indoor monitoring, a wireless sensor network, made up of battery powered electronic devices and motes (microcontrollers equipped with some memory, a radio module, some sensors), has been installed to monitor the quantity of CO_2, CO and dust, the temperature and humidity levels in a specific indoor site (a basement of a building). The testbed was made by 8 nodes placed in the indoor specific site equipped with two ambient boards, featuring CO, CO_2 and dust sensors, in order to collect a number of relevant data about the quality of the air within this area; 2 other nodes equipped with MNA Boards for temperature and humidity sampling were added. Collected data were transmitted to a collection point (i.e. the sink, the eleventh node) at the basement level and covered a period of several weeks. During the experimental phase of the TETRis project, the sink was also connected to a smart object capable of delivering the information collected by the wireless sensor network over the TETRA network, which has been setup in the test bed infrastructure for testing data transmission.

Data collected have been then made accessible in real-time and, and later on, summarized in a report.

At the end of the monitoring process in the specific site, the network can be dismantled and used in a new place (a new building, a new urban area) in order to reduce implementation costs and to repeat the analysis in a different site. The flexibility of wireless sensors networks facilitate the fast and ease deployment of a larger number of sensors.

In the second domain, structural buildings monitoring, another wireless sensor network has been set up for the monitoring of the statics of the same building in order to provide evidence of whether a situation is to be considered "under control" or it is instead critical and, therefore, a prompt intervention is needed and/or some alarm signals need to be sent. The structural monitoring activity was performed by two multi-sensor boards: one was placed indoor and connected to a vibrating wire strain gauge, while the other was placed outdoor and connected to a displacement sensor. Finally, two MNA Boards have added and deployed outdoor to manage wireless connectivity between the basement and the roof of the building used as test bed.

The Fig. 4 shows the test bed area within the Tetris project with the evidence of the sensors location both in the basement and on the roof of the building taken under consideration.

The results of the analysis showed that the MagoNode had a very low current consumption in both transmission, reception and it was also noticed how the MagoNode platform can be easily interfaced to a number of sensors and in particular to the sensors needed for monitoring critical infrastructure and air quality.

The experimental activity demonstrated the efficiency of a wireless sensor network made of an hybrid indoor/outdoor scenario that can be considered relevant both for public safety contexts, in which data about structural and health (air

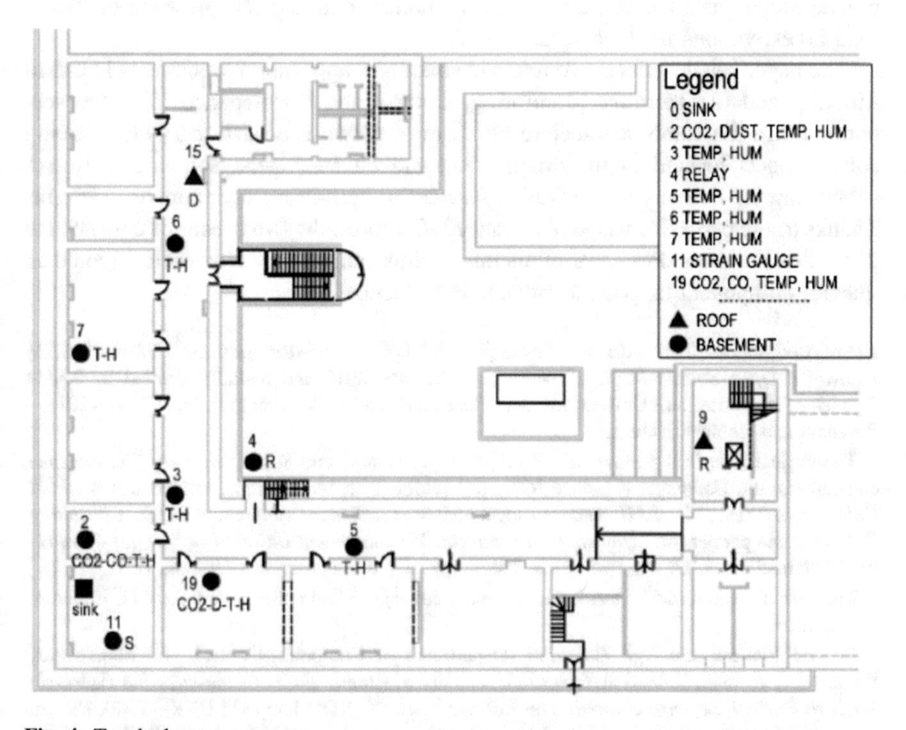

Fig. 4 Test bed area

quality, noise, vibrations etc.) monitoring are made available, and for private needs. Data collected using specific sensors into the TETRis ecosystem have been considered valid for making further analysis and taking the correct actions. The results of the experimental activity confirm that wireless sensor network can be effectively used to support the management of critical situations and to meet the requirements provided by experts on structural health monitoring.

6 Conclusion

We have illustrated some of the activities and results of the TETRis project concerned with the study and experimentation of innovative solutions for the regeneration of urban contexts according to the emerging integrated strategic vision of the Smart City. Experiments were made possible thanks to collaboration agreements with two Southern-Italy municipalities and to the proactive involvement of their representatives in the design and implementation of novel smart city application scenarios in order to provide more effective and efficient services to citizens. A crucial role in the application scenarios is played by the so-called 'social sensors', i.e. the citizens themselves and operators (municipality and utility employees) who are supposed to communicate in real time with all the institutions to report all critical situations such as traffic jams, vandalism and neglect, presence of waste, road holes, various inefficiencies.

The paper has described two relevant smart city application scenarios: (1) Urban Mobility and (2) Territory Monitoring, Control and Maintenance. The two scenarios share the same architecture based on a network of stationary and moving smart objects located in the urban areas and on Intelligent Software platforms supporting the delivery of innovative services to both urban operators and citizens. Thanks to the pervasive usage of advanced ICT tools, the two scenarios enhance the direct line citizens—Public Administration, thus enabling the citizens to become an integral component of good administrative practices.

Acknowledgments This work was developed within the three-years project *TETRis*—"*TETRA Innovative Open Source Services*", started on January 2010 and partially granted by MIUR (Ministry of Education, Universities and Research) under the program PON 2007–2013—Research and Competitiveness.

The project industrial partners are: *Orangee* (coordinator) and *SelexElsag* (two Finmeccanica companies), the *Competence Centre ICT SUD* (along with four member companies: *Methodi, Kaleidos, SIRFIN, SCAI LAB*), and four local SMEs: *H2i, Exeura, Sinapsys and TSC Consulting*. The academic partners are: *University of Calabria, Mediterranean University of Reggio Calabria*, the CNR institute *ICAR* and two Italian inter-university ICT consortia: *CNIT* and *CINI*.

The authors would like to thank the following colleagues and collaborators for their important contribution:

Andrea Vitaletti and *Ugo Colesanti* (Department of Computer, Control, and Management Engineering Antonio Ruberti at Sapienza University of Rome), *Fosca Giannotti, Dino Pedreschi, Barbara Furletti, Lorenzo Gabrielli* and *Roberto Trasarti* (KDD Lab ISTI CNR—Pisa), *Rosario Curia* and *Loredana Sisca* (H2i Srl—Italy), *Michele De Buono* (SCAI LAB Srl), *Raffaele Bianco* and *Salvatore Pirruccio* (Sinapsys Srl—Soverato, CZ), *Roberto De Donato* (SIRFIN SpA—Italy), *Sergio Scrivano* (Methodi Srl—Italy), *Giuseppe Musso, Francesco Scarpelli* and *Luigi Leonetti* (Kaleidos Srl—Cosenza), *Geppino De Rose, Leonardo Acri, Maria Rosaria Mossuto* and *Roberto Caruso* (Municipality of Cosenza), *Luigi Mamone, Corrado Zoccali* and *Vincenzo Settino* (Municipality of Rende).

References

1. Boulton, A., Brunn, S.D., Devriendt, L.: Cyberinfrastructures and "smart" world cities: physical, human, and soft infrastructures. In: Taylor, P., Derudder, B., Hoyler, M., Witlox, F. (eds.) International Handbook of Globalization and World Cities. Cheltenham. Edward Elgar, UK (2011). Available from: http://www.neogeographies.com/documents/cyberinfrastructure_smart_world_cities.pdf
2. Hollands, R.G.: Will the real smart city please stand up? City **12**(3), 303–320 (2008)
3. Chourabi, H., Nam, T., Walker, S., Ramón Gil-García, J., Mellouli, S., Nahon, K., Pardo, T. A., Jochen Scholl, H.: Understanding smart cities: an integrative framework. In: Proceedings of 45th Hawaii International Conference on Systems Science, HICSS-45 2012. Maui, HI, USA, pp. 2289–2297, 4–7 Jan 2012
4. Hall, R.E.: The vision of a smart city. In: Proceedings of the 2nd International Life Extension Technology Workshop. Paris, France, 28 Sept 2000. Available from: http://www.osti.gov/bridge/servlets/purl/773961-oyxp82/webviewable/773961.pdf
5. Moss Kanter, R., Litow, S.S.: Informed and interconnected: a manifesto for smarter cities. Work. Progress 09–141 (2009)
6. Harrison, C., Eckman, B., Hamilton, R., Hartswick, P., Kalagnanam, J., Paraszczak, J., Williams, P. Foundations for smarter cities. IBM J. Res. Dev. **54**(4) (2010)
7. Toppeta, D.: The smart city vision: how innovation and ICT can build smart, "Livable", sustainable cities. The Innovation Knowledge Foundation. Available from: http://www.thinkinnovation.org/file/research/23/en/Toppeta_Report_005_2010.pdf
8. Natural Resources Defense Council. What are smarter cities? Available from: http://smartercities.nrdc.org/about
9. Komninos, N., Schaffers, H., Pallot, M.: Developing a policy roadmap for smart cities and the future internet. In: eChallenges e-2011 Conference (2011)
10. European Commission: Smart cities and communities—support for a better future (2013). http://ec.europa.eu/eip/smartcities/
11. IERC Documents and Publications on the Internet of Things Vision in Europe. http://www.internet-of-things-research.eu/documents.htm
12. Kallman, M., Thalmann, D.: Modeling objects for interaction tasks. In: Proceedings of the Eurographics Workshop on Animation and Simulation. Springer, Berlin, pp. 73–86 (1998)
13. Kortuem, G., Kawsar, F., Sundramoorthy, V, Fitton, D.: Smart objects as building blocks for the internet of things. IEEE Internet Comput. **14**(1), 44–51 (2010)
14. Poslad, S.: Ubiquitous Computing Smart Devices, Smart Environments and Smart Interaction. Wiley, New York (2009)
15. Fortino, G., Guerrieri, A., Russo, W.: Middleware for smart objects and smart environments: overview and comparison. In: Internet of Things Based on Smart Objects: Technology, Middleware and Applications, Springer Series on the Internet of Things: Technology, Communications and Computing (2014)
16. Fortino, G., Guerrieri, A., Russo, W.: Agent-oriented smart objects development. In: Proceedings of 2012 16th IEEE International Conference on Computer Supported Cooperative Work in Design (CSCWD 2012). Wuhan, China, 22–25 May 2012
17. Fortino, G., Guerrieri, A., Lacopo, M., Lucia, M., Russo, W.: An agent-based middleware for cooperating smart objects. In: Corchado, J.M., Bajo, J., Kozlak, J., Pawlewski, P., Molina, J.

M., Julian, V., Silveira, R.A., Unland, R., Giroux, S. (eds.) Highlights on Practical Applications of Agents and Multi-agent Systems, Communications in Computer and Information Science (CCIS), vol. 365, pp. 387–398. Springer, Berlin (2013)
18. Citrigno, S., Graziano, S., Saccà, D.: Smart applications for smart city: a contribution to innovation. In: Proceedings of the Workshops of the EDBT/ICDT 2014 Joint Conference. Athens, Greece, CEUR Workshop Proceedings, vol. 1133, 28 Mar 2014
19. Aloi, G., Bedogni, L., Di Felice, M., Loscri, V., Molinaro, A., Natalizio, E., Pace, P., Ruggeri, G., Trotta, A., Zambia, N.R.: STEM-Net: an evolutionary network architecture for smart and sustainable cities. In: Wiley Transactions on Emerging Technologies, vol. 25, Issue 1 (2014)
20. European Commission: Intelligent transport systems in action, action plan and legal framework for the deployment of intelligent transport systems (ITS) in Europe (2011). http://trove.nla.gov.au/version/166764828
21. Giannotti, F., Nanni, M., Pedreschi, D., Pinelli, F., Renso, C., Rinzivillo, S., Trasarti, R.: Unveiling the complexity of human mobility by querying and mining massive trajectory data. VLDB J. Spec. Issue Paper, vol. 20, Issue 5, pp. 695–719 October 2011
22. PHC09: Pinelli, F., Hou, A., Calabrese, F., Nanni, M., Zegras, C., Ratti, C.: Space and time-dependant bus accessibility: a case study in Rome. In: Proceedings of the 12th International IEEE Conference on Intelligent Transportation Systems (2009)
23. Trasarti, R., Giannotti, F., Nanni, M., Pedreschi, D., Renso, C.: A query language for mobility data mining. Int. J. Data Warehouse. Min. (IJDWM) (2010)
24. Rosi, A., Berti, M., Bicocchi, N., Castelli, G., Mamei, M., Corsini, A., Zambonelli, F.: Landslide monitoring with sensor networks: experiences and lessons learnt from a real-world deployment. Int. J. Wireless Sensor Netw. Seattle (2011). (to appear)
25. Gnawali, O., Fonseca, R., Jamieson, K., Moss, D., Levis, P.: Collection tree protocol. Sen. Sys. 1–14 (2009)

Printed in the United States
By Bookmasters